Hypnofacts 6

Trevor Eddolls

This book is dedicated to

Jill, Katy, Harry, Freddy, Jennifer, Andy, Jake, and Rory

First published in 2018

By iTech-Ed Hypnotherapy

16 Brinkworth Close

Chippenham

Wilts SN14 0TL

Typeset by iTech-Ed Ltd

Contents

Introduction

Like its predecessors, this book also contains various articles for hypnotherapists covering practical issues such as working with clients with anger issues, changing 'bad' habits, and huge section that looks at NLP techniques that can be used by solution-focused hypnotherapists. There are some example word patterns for children and even some thoughts about making your Web site more effective. And there are more theoretical issues around leadership, working with teams, and stress in the workplace.

Again, the articles assume a model of the brain in which core activities (such as telling the heart to beat) are handled by the brain stem, more protective functions (such as fighting, fleeing, feeding, and reproductive behaviour) are handled by the primitive emotional brain, and higher functions (such as problem solving, maintaining attention, and controlling emotional impulses from the primitive brain) are handled by the intellectual brain. In terms of physical parts of the brain, these three areas more-or-less match up to the brain stem and cerebellum, the limbic system, and the cerebral cortex. It also assumes that the primitive emotional brain is very fast and the intellectual brain is much slower and tends to be used less.

In addition, the book assumes that the mind and body make up a single functioning system that is affected by each other and the environment they are in.

And it assumes a solution-focused model for hypnotherapy – moving clients towards their desired goals rather than worrying about the problem itself and its origin.

What's your mindset?

A look at how a client's mindset can affect how they deal with stress.

We've all had clients who seem to deal with issues in life in a really negative way, and, hopefully, we've all had clients who seem happy to deal with whatever life throws at them. What's the fundamental difference between these two types of client? And, perhaps more importantly from the therapy point of view, which one produces clients who are the most stressed?

Back in 2006, Carol Dweck wrote a book called *Mindset: The New Psychology of Success,* which looked at this in some detail. According to Dweck's ideas, some people believe their success is based on innate ability; these people are said to have a 'fixed' theory of intelligence (ie a fixed mindset). Other people believe their success is based on hard work, learning, training, and doggedness; these are said to have a 'growth' or an 'incremental' theory of intelligence (growth mindset). And everyone else is meant to be somewhere on a continuum between the two extremes. And, like most things to do with the mind, people may not be aware of their mindset (ie where they are on the continuum).

So, how can you tell which end of the spectrum a client is if you can't ask them? The answer is that their mindset can still be inferred from their behaviour – especially their reaction to failure. The thinking is that fixed-mindset people dread failure because it is a negative statement on their basic abilities. On the other hand, growth mindset people don't worry about failure as much because they realize their performance can be improved and learning comes from failure.

You've probably come across clients who are right down at the fixed mindset end of the spectrum. They believe their basic abilities, their intelligence, their talents, are just fixed traits, and they only have so much of them. Their goal at work is to look clever all the time and never look stupid. Whereas people at the growth mindset end of the spectrum understand that their talents and abilities can be developed through effort, good teaching, and persistence.

A person with a fixed mindset will tend to avoid challenges, whereas a person with a growth mindset will embrace challenges. When faced with obstacles, fixed mindset people tend to give up easily, whereas growth mindset people will persist in the face of setbacks. Fixed mindset people view effort as fruitless or worse, whereas growth mindset people see effort as the path to mastery. When it comes to receiving criticism, people with a fixed mindset will ignore any useful negative feedback they are given. On the other hand, growth mindset people will learn from the criticism. When other people are successful, fixed mindset people feel threatened by their success, but growth mindset people find lessons and inspiration in the success of others.

As a result of this, fixed mindset people may plateau early and achieve less than their full potential. They have a deterministic view of the world. Growth mindset people reach ever-higher levels of achievement, and this gives them a greater sense of free will. One of the quotes from the book states: "it's not always the people who start out the smartest who end up the smartest."

Dweck concludes that the growth mindset will allow a person to live a less stressful and more successful life.

Although really talking about praising children, Dweck's idea could be extended to clients, when she suggests that you don't say: "good job, you're really clever" because the children are much more likely to develop a fixed mindset. In terms of clients, praising their abilities or achievements will move their mindset towards the fixed mindset end of the spectrum. So what should you say to children (clients)? Try: "good job, you worked very hard". This leads them to develop a growth mindset. It's a good way to encourage clients to persist, despite experiencing some kind of failure, by encouraging them to think about learning in a certain way. So simply praising someone's intelligence harms their motivation and their performance.

Carol Dweck's mindset theory is an interesting way to look at your clients and encourage them to perform as well as they can. And that can lead to successfully achieve their goals.

It's interesting to note that the brain can be affected by the GI tract and *vice versa*. This is called the gut–brain axis (or the microbiota-gut-brain axis). It actually includes the central nervous system, neuroendocrine and neuroimmune systems, including the Hypothalamic-Pituitary-Adrenal axis (HPA axis), sympathetic and parasympathetic arms of the autonomic nervous system, including the enteric nervous system and the vagus nerve, and the gut microbiota.

Gut-brain axis

How does the brain react to bacteria in the gut? The answer is through the vagus nerve and gut bacteria. The vagus nerve connects the brain and the gut. Gut bacteria:

- Break down fibre in the diet into short-chain fatty acids. These can have effects throughout the body.
- Influence the immune system, which may play a role in brain disorders.
- Could be using microRNAs (tiny strips of genetic code) to alter how DNA works in nerve cells.

References:

Carol S Dweck. Mindset: The New Psychology of Success (2007). Ballantine Books. 978-0345472328

More than a bit cross!

Here are a number of ideas to help someone with anger issues.

What can you do for a client who comes in and tells you that they have anger issues? Typically anger is a consequence of stress. It's one of the three ways that the primitive brain tries to protect you against unremitting stress. The answer, for the most part, is to help your client get back into control of their life by emptying their stress bucket and getting back into their intellectual brain. You can also encourage them to have positive thoughts, positive actions, and positive interactions. These are all great things, but what else can you do?

Anger usually results in some form of aggression that can physically and psychologically harm the angry person, other people, or objects in the local environment. Aggression can be physical, verbal, mental, or emotional, or a combination of any or all of the above. There are basically two types of aggression:

- Impulsive aggression (affective aggression), which is unplanned and occurs in the heat of the moment.
- Instrumental aggression (predatory aggression), which is often carefully planned and usually exists as a means to an end. Harming another is the means to achieve a goal, eg a bank robbery.

Anger isn't 'bad', in fact, evolutionarily speaking, it can be a 'good' thing. If a primitive human was being attacked or his food was being taken by other primitive humans, the anger could empower him with a burst of energy to defend himself, his property, and his family. And that could ensure that his genes were passed on to the next generation (as well as that particular behaviour). It's just that anger doesn't go down so well in down town Chippenham.

Anger becomes a problem when:

- A person regularly expresses their anger through unhelpful or destructive behaviours.
- Anger has a negative impact on a person's overall mental and physical health.

And because a person is usually in their primitive brain when they are getting angry, the primitive brain will choose the same neural pathways that it has used before (ie the same habits) no matter whether the outcomes were good or bad last time. It then becomes the job of the prefrontal cortex to overrule the speedy decisions made by the primitive brain. So feelings of anger can be turned down. The issue with teenagers is that their prefrontal cortex is still developing and won't be complete until they are in their early twenties – so they have less of it than most adults. And that can often be infuriating for the adults around them.

Some unhelpful ways people express anger include:

- Outward aggression and violence – such as shouting, swearing, slamming doors, hitting or throwing things and being physically violent or verbally abusive and threatening towards others.

- Inward aggression – such as telling themself that they hate themself, denying themself their basic needs (like food, or things that might make them happy), cutting themself off from the world and self-harming.

- Non-violent or passive aggression – such as refusal to answer questions or ignoring someone (the silent treatment), being sullen, stubbornness, or failure to complete tasks or 'forgetting' to do them.

How angry a client gets and how often they get angry can impact on their health and their mental health. Regular feelings of strong anger or feeling angry for prolonged periods can contribute to illnesses such as gastro-intestinal (digestive) problems, colds and flu, and high blood pressure. Anger can also contribute to mental health problems, making existing problems worse, such as depression, anxiety, eating problems, or self-harm. And it can contribute to sleep problems, and problems with alcohol and substance misuse. Anger issues can often be found in people with mental health issues such as Borderline Personality Disorder (BPD), other personality disorders, psychosis, or paranoia.

We've said that stress can cause long-term feelings of anger, but what sorts of things can trigger an angry outburst? Anger usually occurs when a person feels:

- Threatened or attacked
- Frustrated or powerless
- Unfairly treated.

Obviously, not every time a person feels like that leads to angry outbursts. And the same event may not lead to one of the feelings listed above. Instead, a person may just feel amused by what's happened, or they may just feel annoyed rather than angry, or they may feel something else (or nothing at all).

We'll be bucket emptying and relaxing a client in the hypnotherapy sessions, but what else can we suggest to a person to do to help control their anger? Firstly, like so much else in life, we can discuss their sleep pattern, their eating habits, and how much exercise they are getting. Often, lack of control goes with a sleep deficit and we can help a client to get the right amount of sleep for them. We can make general suggestions about what they are eating and how much. And we can usually suggest more exercise – a brisk 30-minute walk usually has benefits, including making people more ready for sleep at the right time. More intensive exercise can help reduce the amount of cortisol in the blood and that reduces feelings of anger.

Clients will often come up with other lifestyle changes that will help them to remove from their life whatever it is that's causing them so much stress, eg to cut down on their drinking because that's when they get angry the most, or to change the places they visit so they're not somewhere that they often get angry, etc.

What else can people do? A suggestion from CBT is that they recognize triggers and think about what else the trigger might mean. It might not be that the other person is criticising them personally, they may be criticising something more abstract. The client could ask themselves why the other person is saying these things. What might have happened in that other person's life that is making them act in this way. Using this technique, the client can become disassociated from the situation. The other thing to do when recognizing a trigger is move away from the situation – to go somewhere that the trigger situation isn't. And that way, they won't get angry. Even simply counting to 10 can give back control to a person, so that they can choose how they want to respond to a situation. And a suitable response may well take the form: "when you do xxx, I feel angry".

> Researchers at the Karolinska Institutet in Stockholm, Sweden identified neurons that drive aggression. In a study on rats, they found that the rodents displaying the highest level of aggression also had more active neurons in the ventral premammillary nucleus (PMv) in the hypothalamus.

Again from CBT, there are certain phrases that the potentially angry person should avoid saying because they will make the person they are talking to angry, and that leads to an unhappy vicious circle. The phrases to avoid include:

- Always, eg "you always do that".
- Never, eg "you never listen to me".
- Should or shouldn't, eg "you should do what I want", or "you shouldn't be on the roads".
- Must or mustn't, eg "I must be on time", or "I mustn't be late".
- Ought or oughtn't, eg "people ought to get out of my way".
- Not fair.

Other suggestions for dealing with anger include:

- Breathing slowly and breathing out for longer than spent breathing in (7-11 breathing).
- Distraction – thinking of something else can help stop feelings of anger increasing. This could be listening to music or going for a shower.
- Relaxation – people are less likely to get angry when they're relaxed. Things like yoga and mindfulness can help with relaxation.
- Forgiveness – stops feelings of anger filling a person's thoughts. It allows them to move on with their life.
- Humour – if possible, humour can diffuse a situation that might well lead to anger.

Clearly, getting a client back into their intellectual brain will give them back control of their life. Getting them to recognize the stress in their life and the angry habits they have developed and are using can help them to change. Some of the ideas in this article can help them to see positive change more quickly.

References:

https://www.verywellmind.com/what-is-aggression-2794818

https://www.livestrong.com/article/258737-aggressive-behavior-in-adults/

https://www.mind.org.uk/information-support/types-of-mental-health-problems/anger/#.Wt4rqpch1PY

https://www.mind.org.uk/information-support/types-of-mental-health-problems/anger/causes-of-anger/#.Wt4vIJch1PY

https://www.mind.org.uk/information-support/types-of-mental-health-problems/anger/managing-outbursts/#.Wt4vR5ch1PY

https://www.mind.org.uk/information-support/types-of-mental-health-problems/anger/long-term-coping/#.Wt4vdZch1PY

https://www.mayoclinic.org/healthy-lifestyle/adult-health/in-depth/anger-management/art-20045434

https://www.nhs.uk/conditions/stress-anxiety-depression/controlling-anger/#dealing-with-anger

https://www.medicalnewstoday.com/articles/321941.php

Let me sleep

A brief look at dealing with a partner's snoring.

We all need sleep. And if we don't get enough, we can actually perform less well at all sorts of tasks the next day. It helps if we get up at much the same time every day and go to bed at much the same time every evening. And after a good night's sleep, we feel energized and ready for anything.

Your brain is really quite clever. It can do lots of things for us, without us giving it a second thought. For example, if we go into a room that has a particular perfume smell, after a while, we stop noticing it. Or, if we live by a particularly busy road, after a little while, we stop noticing the noise from the cars. This process has a name, it's called habituation. It simply means that our brain can pretty much get used to anything if it happens for long enough – and all without us giving it any conscious thought. There are even stories of people being able to hold normal conversations in noisy factories because their brains have stopped hearing the noise – and yet any newcomer is almost deafened by the sound. Or you may have been to a party and heard your own name being said even though there is lots of party noise. Or you may have found yourself engrossed in conversation with someone at a party, and you've both stopped hearing all those other party sounds. Your attention is on what's being said and your brain has habituated the other party noise and you don't register any of it. It's perfectly natural.

And your brain generally starts turning off sounds as you start to drift off to sleep. The sounds might still be there, but they get ignored or sometimes what's called the auto-symbolic effect takes place, where the external sound will be changed by your unconscious mind to some other sound and that might well be incorporated into your dream. Your brain's ability to get used to any noise means that you can go to sleep on a train, on the sofa with the TV on, at the cinema, just about anywhere. You know that you don't need silence in order to get to sleep.

So, why isn't this happening when you can hear someone snoring? Why does someone else's snoring make you feel exhausted or angry or even depressed? I'm afraid it's your brain that's to blame – again!

At first, you may well have gone to sleep with the sound of snoring going on near you, but once you've been kept awake by the snoring sounds, you begin to anticipate being kept awake next time. You expect the snoring to keep you awake and that starts to be stressful – and that's long before the first snoring sounds begin. Your body begins to listen out for the snoring sounds that are going to keep you awake. Your body, instead of letting itself drift naturally into sleep, is on full alert waiting and waiting for the snoring that you know will come and you know will keep you awake. A self-fulfilling prophecy. You predict that the snoring will keep you awake and you lie there awake until it does!

So, what can you do? Once you understand what's happening in your mind, you can let yourself relax. You stop monitoring what's going on until the snoring starts. You can let the natural habituation take effect. You can switch of the sounds, you can switch off the monitoring and alertness, and you can simply drift off to sleep as you would on any other occasion – in front of the TV or wherever.

You give yourself permission to let nature take its course and drift off to sleep. And wave goodbye to any feeling of anger that you have been kept awake. And welcome the fact that your partner can drift off so quickly and easily because they feel so comfortable in your presence. And you feel pleased that you can make someone feel that way. And you just relax, turn off your day-time alertness, and let the natural effects of habituation allow your brain to ignore the sound completely. And you get a really good night's sleep.

The truth is that people can get used to any particular noise. People can sleep near busy roads and don't 'hear' the sound after a while. Or they can sleep in a farmhouse and not be disturbed by the sounds of sheep or cattle, etc. This process happens so often, it has many different names. It is called habituation or acclimatizing or accustoming or getting used to. And people can habituate to the sound of their partner sleeping.

Paradox – changing habits the Ericksonian way

Here's how Milton Erickson would change clients' habits.

Bad habits, like smoking or waking up in the night feeling anxious, are some of the things that bring clients to see us. Milton Erickson had a unique way of dealing with these kinds of condition. He would disrupt the behaviour His thinking was that emotions cause problem behaviours, and, in turn, those problem behaviours affect emotions. Therefore, if an appropriate task can be performed by a client, it can begin to disrupt the habit or behaviour. And once the behaviour has changed, it will start to change how a client feels.

One well-known story is how Erickson treated a client for insomnia. Because the client hated polishing the wooden floors in his house, Erickson set him a task of polishing all his wooden floors whenever he was still awake twenty minutes after going to bed. If the client became sleepy while polishing their wooden floors they could return to bed. But if they didn't fall asleep within twenty minutes, they had to get up again and carry on polishing.

The result of this was that the client very quickly started drifting off within minutes. His unconscious motivation to sleep had become as strong as his conscious motivation. This is an example of a slightly 'painful' task being set that links to the problem behaviour and, in turn, makes continuing the behaviour difficult. It works best if the task set is one that needs to be done, but that the client has been putting off.

Similarly, for a client wanting to stop smoking, Erickson might explain that they could smoke, but they needed to buy one cigarette at a time and walk to the shop each time they wanted a smoke. This both encouraged healthier behaviour and altered the typical behavioural pattern. Usually clients got tired of the habit fairly quickly and stopped on their own.

Another case that Milton Erickson dealt with was a chronic alcoholic. The task that was set for him was to go to the local Botanical Gardens and visit the Cactus House. Once there, he was to stay for at least an hour and during his visit 'do a lot of thinking'. Clearly, cacti do not need much water – metaphorically, they can go for a long time without a drink. The client 'learned to respect' cacti and stopped drinking. Interestingly, his wife, who was also a heavy drinker, gave up drinking as well. This kind of metaphorical task can indirectly illustrate a solution to a client's problem and generally speed up progress.

Another technique is to tell the client that their task is to perform their 'bad' habit. This is called prescribing the symptom and it can be very effective because it creates a paradox – an 'uncontrollable' compulsion becomes a chore that they would rather not do.

A fourth technique is to change the steps involved in a pattern of behaviour (a pattern interrupt). Typically, problem behaviours become rigid and fixed, so introducing a new element can change the whole pattern and make it harder to maintain. That's the theory – in practice, the client can change:

- Frequency/rate
- Duration
- Time of day
- Location in the body
- Intensity
- Quality
- Sequence
- Interrupting
- Jumping to the end
- Adding or subtracting elements
- Breaking up large pieces of a sequence into small pieces
- Performing the symptom without the pattern
- Performing the pattern without the symptom
- Linking the pattern to another (usually undesired) pattern (called symptom-contingent tasks by Erickson).

Although it seems that we're giving them permission to continue their 'bad' habit, by changing part of the pattern, the whole thing can stop. For example, a nail-biter might agree to only bite their nails at a prescribed time.

Similarly, with a fear, you can find out in what order the client experiences the symptoms, then have them imagine doing what they are afraid of and having the symptoms occur in a different order. And this works well with weight loss. Rather than losing weight being a chore, then hitting their target, and thirdly binging as a treat, and starting their yo-yo diet again Erickson told one client to gain a specific amount of weight. That meant she hit her target and then she could lose weight. Gaining the weight then became the chore and the client stayed at their lower weight. The usual pattern had been flipped over.

Erickson would always build rapport with his clients and get a commitment from them before revealing the task, Erickson would often tell his clients that he could solve their problem but they weren't going to like it. He would only reveal the task after the client begged for the task and promised they would do as he said.

It could be a useful technique to try with clients.

References:

Ericksonian Hypnosis: Breaking Habits with Tasks: http://www.hypnosis101.com/hypnosis/ericksonian-hypnosis/habits-tasks/

Using Tasks to Disrupt Problem Patterns: https://www.unk.com/blog/4-useful-therapeutic-tasks-types/

New Year's resolutions

How you can help your clients keep their resolutions.

Too many people decide that they want to make a change in their lifestyle and use the New Year as a way of making the required change. Or that's what they plan to do. And sometime in January, too many people find that they haven't succeeded in giving up alcohol, losing weight, stopping smoking, going to the gym, or whatever, and go back to their old lifestyle once more. So, how can people make and keep their resolutions?

Firstly, many resolutions fail because:

- They are based on what someone else (or society) is telling you to change.
- They are too vague.
- You don't have a realistic plan for achieving your resolution.

The goal should be SMART (Specific, Measurable, Achievable, Relevant, and Time-bound).

There's evidence to suggest that a person's chances of success are greater when they channel their energy into changing just one aspect of their behaviour. So, it's recommended that people make only one resolution rather than lots.

It seems that humans are driven by 'loss aversion', ie people are more motivated to recover loss than they are to win gains. So, resolutions should be worded to recover something lost, eg an old hobby or a former level of fitness. They also must be realistic. People are more likely to keep resolutions if they can see them as being somehow important to other people, according to Dr John Michael, a philosopher at Warwick University. Making resolutions public can help people keep them because the fear that people will think worse of them if they don't see them through adds to their resolve.

It's also important to plan for what you want to achieve, identify any obstacles that you'll meet, and identify ways round them. Charles Duhigg, author of *The Power of Habit,* suggests thinking of New Year resolutions as New Year plans. He suggests that rather than setting a far off goal, eg running a marathon, it's better to set an immediate plan that you can start straight away. So your marathon goal might begin with the goal of running half a mile every Monday morning, and building on that.

Duhigg suggests breaking down a new habit into its three parts: a cue, a routine, and a reward. For running, a cue could be just putting on a person's running kit, even if, to begin with, they don't go running. And then they get a reward, which helps their brain to establish the behaviour. These small steps can then build up to running a marathon.

Implementation intentions is a technique that uses an 'if-then' structure. So a resolution might be to run half a mile on Monday mornings. The implementation intention could be: "If it's Sunday night, then I will set my alarm 30 minutes earlier, so that I have time to run". The rule is to identify the situations related to the cue in order to find the 'ifs' and link them to appropriate responses to make the 'thens'. A recent study by Chris Armitage, professor of health psychology at the University of Manchester, found that 15% of smokers who formed implementations stopped smoking, compared with 2% of those who did not.

One of the obstacles that people face, for example with running a marathon, is that running a mile may be OK, but they still have to run 25 miles more. A 2012 study published in *The Journal of Consumer Research* found that focusing on the smaller number in reaching a goal kept people more motivated. So, instead of looking at the big number left to get to a goal, look at what's already been achieved. Later on, when that goal number is much smaller, focus on what little remains to achieve the goal.

It's interesting to note that a study by Marion Fournier, a lecturer at the Université Nice Sophia Antipolis, found simple habits form more quickly in the morning than in the evening. Researchers believe this may be to do with levels of the stress hormone cortisol, which tend to be highest when we wake up. Apparently, cortisol elevation changes the mechanisms in our brain, blocking the prefrontal cortex, resulting in a behaviour becoming habitual.

Gabriele Oettingen, professor of psychology at New York University and author of *Rethinking Positive Thinking: Inside the New Science of Motivation*, suggests that people shouldn't daydream about their future success because they'll have less actual success. She suggests that it's better to look at what obstacles are in the way and how to get over them – Oettingen calls this technique WOOP (Wish, Outcome, Obstacle, Plan):

- Wish – what do you want?
- Outcome – what would the ideal outcome be? What will your life look like when you hit your goal?
- Obstacle – you know yourself. What will try to stop you? What has sidelined you before?
- Plan – how will you get around it?

Similarly, Gretchen Rubin, the author of *Better than Before*, suggests it's crucial to avoid listening to the excuses that make our habits falter, such as the false choice loophole, eg you can't go for a run tomorrow because you have to do X. Recognizing them in advance can make them less powerful, when you realize you're doing it, you're much more likely to resist.

And should your New Year plan be flexible or rigid? In a 2015 study, researchers paid two groups of people to go to the gym for a month. Group 1 was paid if they started a workout within a two-hour window they chose in advance. The second group was paid whenever they went to the gym. The result after a month was that group 2 was more likely to stick with the gym habit. So be flexible with your new habit.

In contrast, Prof Neil Levy at the University of Oxford suggests making detailed resolutions is important, for example: "I'll go to the gym on Tuesday afternoons and Saturday mornings" is more likely to be successful than simply saying "I'll go to the gym more".

And treat everything like an experiment. If something doesn't work, then treat that as more data for what will work. Remember Edison took 200 (in some versions 1000) attempts to develop a working light bulb. Treat any failure as a temporary setback rather than a reason to give up altogether.

For a New Year's resolution to be successful, it needs to be as easy as possible. A study showed that people who travelled 8km to the gym went once a month, whereas people who travelled 6km went five or more times a month. "That 2km makes the difference between having a good exercise habit and not. That is how our habitual mind works – it has to be easy.

And be kind to yourself. For many people, according to Dr Jessamy Hibberd, a clinical psychologist, the biggest obstacle to new habits is self-criticism. Study after study shows that self-criticism is correlated with less motivation and worse self-control, in contrast with being kind or supportive to yourself, as you would to a friend – especially when confronted with failure.

Whatever you set as your goal, solution-focused hypnotherapy can help you to achieve it.

References:

https://www.theguardian.com/lifeandstyle/2017/dec/29/experts-guide-making-keeping-your-new-year-resolutions

https://www.nytimes.com/guides/smarterliving/resolution-ideas

http://www.bbc.co.uk/news/uk-42353226

https://www.nhs.uk/Livewell/Healthychristmas/Pages/NewYearresolutions.aspx

Using poems as therapy

Investigating the value of poetry in therapy.

Clearly, the work we do helping people to relax and get into their logical intellectual brain and empty their stress buckets works well with most clients. And between sessions, clients are expected to listen to our audio tracks (on CD or as downloads). But what if we could add something else to help them? Nothing that would take anything away from our core activities, but something that would help them process their feelings. That's what poetry therapy can do.

Poetry therapy (or almost any form of concise writing) can be used with clients for any number of reasons, including:

- Understanding themselves better
- Healing from the profound wounds
- Giving a voice to thoughts and emotions that have been difficult to articulate
- Developing their self-esteem
- Learning to express themselves
- Gaining a more realistic perception of themselves
- Developing greater empathy
- Releasing emotional pain
- Alleviating anxiety and stress
- Discovering new meaning, purpose, and passion in life.

And the benefits of poetry therapy include:

- The opportunity to reflect on one's life
- A new sense of hope when the future seems bleak or hopeless
- Greater self-confidence, especially in terms of trying new things
- Increased ability to forgive hurts and let go of regrets
- Greater insight into oneself
- Finding comfort and healing following a painful loss
- Healing from trauma
- Ability to look at life or specific events from a new perspective
- Improved coping and problem solving skills
- Decrease in symptoms of depression and anxiety
- Courage to face challenges and overcome set-backs
- Cognitive stimulation
- Improved interpersonal skills

- Stronger and more effective communication skills
- Greater ability to self-soothe and handle stress
- Healthier and more satisfying relationships with others.

You may find that clients want to explore feelings and memories, and poetry can help with this because it can:

- Be used as a vehicle for the expression of emotions that might otherwise be difficult to express.
- Promote self-reflection and exploration, increasing self-awareness, and helping individuals make sense of their world.
- Help individuals redefine their situation by opening up new ways of perceiving reality (a reframe).
- Validate emotional experiences and improve group cohesiveness by helping people realize many of their experiences are shared by others.
- Help therapists gain deeper insight into those they are treating

In general, poems can be concise, address universal emotions or experiences, offer some degree of hope, and contain plain language. Some poems that are used include:

- The Road Not Taken by Robert Frost
- The Journey by Mary Oliver
- Talking to Grief by Denise Levertov
- The Armful by Robert Frost
- I Wandered Lonely as a Cloud by William Wordsworth

Obviously, other poems may be appropriate for a particular client.

No-one really knows how any therapy works, but poetry therapy often follows the model devised by Nicholas Mazza. In this model, poetry therapy has three components:

- Receptive/prescriptive
- Expressive/creative
- Symbolic/ceremonial.

The receptive/prescriptive part sees the therapist introduce a relevant poem and encourage the client to react. The poem is usually read aloud by either the therapist or the client. While the poem is being read, the therapist notes the verbal and nonverbal reactions of the individual, which leads to follow-up questions such as, "I noticed you were smiling as the poem was being read. Can you tell me about your reaction?" "Is there a particular line in the poem that touched you?" "How does this poem make you feel?"

The expressive/creative part involves creative writing – poetry, letters, and journal entries. The process of writing can be both cathartic and empowering, freeing blocked emotions, and giving voice to a client's concerns and strengths.

The symbolic/ceremonial part involves the use of metaphors, storytelling, and rituals as tools for effecting change. Metaphors can help clients explain complex emotions and experiences. Rituals may be particularly effective to help those who have experienced a loss to address their feelings. Writing and then burning a letter to someone who died suddenly, for example, may be a helpful step in the process of accepting and coping with grief.

There are a variety of organizations that train people to become poetry therapists. I'm not suggesting that we need to go that far. What I am suggesting is that giving clients a poem to read can help them to not feel so alone with their problem, and can help them to express how they feel. I wouldn't ask a client to write a poem or letter, but I certainly wouldn't stop them because of the therapeutic effect of expressing their thoughts.

I have used poems from time to time with clients and they have responded well – particularly clients with grief issues. I've used Do Not Stand At My Grave And Weep by Mary Elizabeth Frye and Funeral Blues by W H Auden with them.

It's a technique that's worth trying with some clients, I think.

References:

https://www.goodtherapy.org/learn-about-therapy/types/poetry-therapy

https://www.psychologytoday.com/blog/minding-the-body/201101/will-poem-day-keep-the-doctor-away

https://www.addiction.com/a-z/poetry-therapy/

Scripts to use with children

Tides

If you've ever been on holiday at the beach, you'll know that sometimes the tide comes in – and there's less beach for people to sit on and play. And sometimes the tide goes out – and you discover more about the beach. Perhaps you find rock pools with small sea creatures living in them, which you hadn't seen when the tide was in. There's usually more sand, and people spread out more, and there's more room to play and more places to investigate.

Your brain is a bit like the beach in some ways! If you look at a photo of a familiar beach taken 20 years ago, you still recognize it as the same beach – although the clothes people are wearing look slightly odd, and the cars in the car park seem different. But the beach is the same.

Your brain likes to keep everything the same. So if your brain has behaved in a particular way once, it will try to behave in the same way every time – even if, on reflection, that wasn't the best way to behave. So when the tide comes in, it always goes up to pretty much the same height (where those few stones and dried out pieces of seaweed seem to have been left).

And you might wonder, as the sun tells you it's early afternoon, whether the beach could dig a big trench for the seawater to go in – and then all the children and parents would have much more beach to play on. That would be great.

But obviously the beach can't dig a trench.

But perhaps your brain might make a positive change. Let's supposed that you're confronted with the same situation every day (just like the tide coming in and going out). For you, for example, you have to learn some spellings – or 'words' as some people call them. Or you have to learn a particular times table. Or, it might be that someone asks you to empty a bin.

It can be so easy to behave the way you did yesterday, and the way you did the day before, and the way you did the day before that, going back to your last birthday, and the one before that, and the birthday before that, and perhaps even the one before that. The tide has come in and gone out again, and the ten year old you has acted like a six year old version of yourself! And you're probably asking yourself, is that what you want to happen?

And that's where your intellectual brain comes in. It is logical and sensible. And when you're relaxed, your intellectual brain can take over control – and you behave in ways that seems sensible to the 10-year-old you.

And gradually, as you behave in those new grown-up ways, those old ways of behaving disappear. And you feel much happier and in control of your own life.

And now I want you to just picture yourself so nicely relaxed ... letting yourself feel as calm as you've ever felt. And remembering how the intellectual brain of the 10-year-old you can decide on the best way of behaving. You don't have to do what you did yesterday, or the day before, or when you were 9 or 8 or 7 or 6 – an incredibly young version of you.

And whenever you need to, you can remember or imagine these wonderful feelings of relaxation and let your intellectual brain make logical decisions for how you want to behave when you're 11, or 12, or even 13.

Now, I'm going to count from 1 to 3 and you'll be awake feeling incredibly relaxed.

1   your whole body is beginning to feel awake and alert

2   your eyes are feeling sparklingly clear.

3   you feel full of energy and nicely calm and relaxed. Open your eyes, take a good deep breath, and stretch.

Spaceship

And while you're lying there so nicely relaxed and calm, I want you to imagine that you are on a spaceship ... a nice big comfortable spaceship that's miles from the Earth, and still has miles to go to a new space colony. But everything is working perfectly except for the robots, which can't come online until you are nearly at your destination.

You are the captain and you have four crew members to help you. You have to make sure that there is plenty of air to breathe – but not too much. There has to be just enough food each day, but not too much waste. The spaceship has to be kept tidy so that things can be found in an emergency. And the recycling has to be put in the recycler each day so that nothing is wasted on this long journey.

There are five of you on duty, so you decide to divide the jobs up equally between the five people. Everything seems to be OK until one evening when you go for your dinner, you find there isn't any. One of the crew didn't do their job. The next day, you find you have no clothes to wear. One of the crew didn't put on the washing. And the next day, it's very hard to breathe. One of the crew forgot to recycle the air.

As captain, you call everyone to a meeting. Is it fair if only four people do their tasks and not five? Is it fair if only three people do all the tasks or just one person? Is it fair if you have no food, no clean clothes, no air to breathe?

Everyone agrees with the captain. They will do their jobs, and life will be better for everybody.

The captain takes a deep breath – he knew they were a great crew.

And now I want you to just picture yourself so nicely relaxed on that couch, letting the very last of those feelings of anger and frustration simply disappear up to the sky. Letting yourself feel as calm as you've ever felt. And remembering how you were the captain of the space ship, and how you and your crew all had to do their jobs in order for the spaceship to reach its destination.

And just remember that whenever you need to feel relaxed, you can just picture yourself on that couch feeling so amazingly calm.

Hll forts

Every New Year, lots of people make New Year's resolutions. The set themselves simple goals that they want to achieve. They might decide to go to the gym, or go jogging, or they might decide to spend more time with their family – all sorts of different things.

But you don't have to wait until the end of one year before you can set yourself targets for the following year. You can do it any time you like. You might decide that you want to be better at football. Perhaps you set yourself a target of keeping the ball in the air for 21 kicks or hits with your leg. At first, you might find that you can only do 2 or 3. Then, after you keep practising, you can do 5, and then 7. And then one day, you manage to get through the 10 barrier. You can keep that ball in the air for 11 different kicks or other leg contacts.

And the next time you try, you can only do it, for, perhaps, 3 goes. And you could give up then, You could decide you want to try something else instead. But you know, that every new thing you start, you won't be brilliant at, until you've spent lots of time practising. And if you keep on with the football, you'll find one day you get to 16 successful attempts. And as time goes by, you'll get to 21 kicks. And the more you practice, the more often you'll be able to do 21 keepyuppies – or more than 21.

And that's the same for anything else you want to learn in life. The more time you spend trying to achieve your goals, the better you will get, and you'll be so pleased at how good you get – simply by practising. And the other thing you'll be really pleased to discover is that the more you practice, the easier it becomes to get better. It's like starting to run up a very steep hill to a hill fort. Hill forts are often built on hills shaped like a semi-circle. You'll find lots of people will just give up. But if you keep on, the slope gets flatter and flatter until when you're very near the top, when you're very close to your goal, you'll be able to get better so easily. And you'll be amazed by the others who are being put off by the very steep sides near the base of the hill, struggling to achieve a new goal for them.

And sometimes, in your mind, you think of things that you'd really like to achieve – perhaps you'd like to learn to play the guitar, perhaps you've always wanted to learn how to sail a boat – who knows. And you can think of new things every day. And in your mind, each of these goals might look like steep-sided hill, but you know that the more you struggle up the hill, the flatter the sides become and the easier it becomes to reach the hill fort at the top.

And the great thing is that other things that perhaps your teacher or your mum and dad tell you to do fit the pattern exactly. To begin with, they can seem really hard. It's like climbing up a steep-sided hill with a fort at the top. And you just feel that it's too hard to continue and you want to give up. But, like everything else, if you carry on pushing up that hillside, you'll notice that the sides start to get flatter and things start to get easier too.

And while it isn't easy to achieve a goal, it does get easier. And you can be so pleased with yourself as you find yourself getting closer and closer to your target. And you find things getting easier and easier. And you'll be so pleased with yourself as you achieve the different goals you set yourself.

What works?

Hypnotherapists sometimes use other techniques to help their clients. Here's a look at the pros and cons of power posing and affirmations.

Poser poses

It all started with some research by Amy Cuddy and colleagues from Harvard University, who studied body language and the impact it has on your hormones. They classified different body positions as 'high power' or 'low power' poses. As a general rule, the high power poses are open and relaxed (eg stand up tall, with your feet spread, and your hands on your hips), and the low power poses are closed and guarded.

In the study, they took a saliva sample from each subject (there were 42) and their testosterone and cortisol levels were measured. Next the subjects were asked to sit in either a high power pose or a low power pose for two minutes. And then a second saliva sample was taken to measure their testosterone and cortisol levels. What they found was that high power poses increased testosterone by 20 percent and decreased cortisol levels by 25 percent. Other research has shown that powerful leaders tend to have higher levels of testosterone and lower levels of cortisol. And those two hormone levels are important because people (men and women) with higher levels of testosterone generally have increased feelings of confidence, and people with lower levels of cortisol tend to be less anxious and are better at dealing with stress.

So all you need to do, before your client is doing anything that makes them feel anxious, is get them to spend a couple of minutes adopting one of those power poses, where their posture is very open, they're standing tall, and looking up. And that will help them to feel more confident and relaxed, and generally get them in the mood to deal with anything.

Or that's what everyone thought. Now, the thing about science is that experiments are meant to be repeatable, ie if I do exactly the same experiment that you did, I should get pretty much the same results that you did. So some researchers started to recreate the original experiment. Nine experiments were performed and none of them found the same results as the original. And that's a result that's similar to the cold fusion experiments that were all the rage back in the early 1990s. No-one could recreate the same results.

So is that the end of the idea of power posing? Well, yes and no. There are a couple non-scientific phrases that you hear therapists use. They are, "act as if" and "fake it till you make it". Both are designed to encourage people to achieve their goals. So, let's suppose you are a newly-qualified teacher in a new school, with children whose names etc you don't know yet, how do you behave? Well you might feel a little shy because of your newness and behave like a shy person would. But that's not a very effective way to start your teaching career in a school. Because you know how a teacher is meant to behave, you can 'act as if' you are that experienced teacher. And you can fake being an experienced teacher until you make it – until you become that experienced teacher. And that seems to work in a lot of situations that aren't connected to schools.

So, what I'm suggesting is that power poses may not increase your testosterone levels or decrease you cortisol levels, but what they might do is make you feel more powerful going into a situation. And, remember, that how people act towards you is based on how they perceive you – what they infer from your behaviour. So, if they see you looking confident, relaxed, and not stressed, that's how they are likely to treat you. And that feeds back to you and how you act with them. So it may be worthwhile giving it a go with clients, just don't expect their hormone levels to change.

Affirmations

Affirmations (you might think of them as mantras) are basically a form of auto-suggestion, where a person repeats certain phrases, and feels better about themselves and their ability to face whatever the day has in store for them. An example would be something like: "I expect to be successful in whatever I have to do". The word 'affirmation' comes from the Latin 'affirmare', which means to make steady or strengthen.

So how do they work? The idea behind them is that we are what we think. If you expect to be able to solve every problem that life throws at you (even if it does involve a bit of googling), then that's what you tell yourself when a new problem emerges. And that's pretty much how you act in that situation. If you tell yourself that the life is far too complicated for you to get your head around and you're going to need help, then that's what happens. And there's a physical neuroscience-based reason for this. It's all based on neuroplasticity. What that means is that the neurons in your brain can move around to where they are needed. And these neurons then connect together more strongly. At the end of a neuron are tiny dendrites (short extensions), and these can connect (across a synapse) to others. The more connections there are and the more neurons that are connected affects how strongly a behaviour or habit is. And you create these habits by doing something over and over again – such as reading a manual until you know it thoroughly. In fact, it's how you learned to drive so that you can now do it without giving it a second thought – unless someone pulls out in front of you!

Our behaviour and our thought patterns all become embedded by this kind of repetition – that's what makes establishing new habits so hard to begin with. So constant repetition of these affirmations is meant to gradually create more connections and become fixed as established ways of thinking. And so you start to think about yourself more positively. The idea is that each week you choose one or two affirmations and say them out loud (or you could write them down) several times a day. You can make affirmations up yourself, depending on what your particular need is at the time (eg eating less, going to the gym, staying calm around children, etc). Make sure that they are written in the present tense because the phrase is meant to represent how things currently are (and soon that's how they will be).

Émile Coué in the early 1900s gave us: "Every day, in every way, I'm getting better and better". Other examples you could use are:

- I breathe in relaxation. I breathe out stress.
- Even when there is chaos around me, I remain calm and centred.

- I am free of anxiety, and a calm inner peace fills my mind and body.
- All is well in my world. I am calm, happy, and content.

Like all quick fixes and things that seem too good to be true, there is evidence that affirmations don't work quite as well as everyone claims. For example, a study in 2009 found that positive affirmation had only a small, positive effect on people with high self-esteem. However, affirmations had a detrimental effect on people with low self-esteem. In addition, people with low self-esteem who made positive affirmations felt worse than individuals who made positive statements but were allowed to consider ways in which the statements were false. As an alternative, some studies found that self-affirmations involving writing about a person's core values could improve performance under stress.

So, if you think that your client's self-esteem is good enough, these affirmations, at worse, won't do them any harm. And, who know, they may make things better.

References:

https://www.inc.com/business-insider/amy-cuddy-the-poses-that-will-boost-your-confidence.html

http://jamesclear.com/body-language-how-to-be-confident

http://www.newsweek.com/power-poses-dont-make-you-more-powerful-studies-664261

http://fortune.com/2016/10/02/power-poses-research-false/

http://liveboldandbloom.com/09/quotes/positive-affirmations

https://en.wikipedia.org/wiki/%C3%89mile_Cou%C3%A9

http://www.huffingtonpost.com/dr-carmen-harra/affirmations_b_3527028.html

https://en.wikipedia.org/wiki/Affirmations_(New_Age)

I'll do it in a minute!

Some thoughts on procrastination – if you can be bothered to read them!

One of the most frustrating things when working with clients is to set them some 'homework' (experiment or whatever you call it), and find they haven't done it, even though they've had 168 hours, and they did agree to do it by the time you next met. The truth is that we've all had times when we've procrastinated, but some people seem more prone to put things off to some unspecified future time than others. So let's suppose it is you who is procrastinating – what's going on and what can you do to get back in control of your work and your life?

Let's do some neuroscience. As you know, the brain is divided up into areas that have names. One of those areas is the prefrontal cortex, which is the last part of the brain to develop and isn't fully formed until people are in their early 20s. This is like mission control for a person and makes sensible logical choices. It looks at long-term goals and tries to override choices made by other parts of the brain. That's the part saying that you really need to do your homework (or whatever) now. It's like your internal grown-up.

The dorsal striatum is an older part of the brain (in evolutionary terms) and deals with cognition involving motor functions and is associated with the brain's reward system. It is associated with motor actions that result in some kind of reward. This is used in Pavlovian conditioned reflexes. The trouble is that most of the time it suggests that you do what you've always done – which is probably to check your e-mail and social media first before you do any hard work. These actions therefore put off doing real work until later – often much later.

The nucleus accumbens is part of the ventral striatum (again an older brain part). It plays an important role in processing rewarding stimuli, reinforcing stimuli (eg food and water), and those things that are both rewarding and reinforcing (exercise, drugs, etc). And this brain part really is looking for the fun choice of what to do. And that means it will be suggesting getting a drink and a snack before starting work and then checking social media, and then sharing that funny cat video with your friends, and then, well, just about anything else that isn't work.

Typically, the prefrontal cortex is able to keep control of the situation and make the final decision, but sometimes it isn't. And one of the most common reasons that the prefrontal cortex loses control is stress. So how can you reduce stress? Here are some dos and don'ts. The first thing to do is separate work time from home time. Don't spend the evening checking e-mails or doing some extra work to catch up. Similarly, don't work overtime. Both of those increase stress. Walking in the countryside and listening to your favourite music are good ways of reducing stress. You can also set aside a specific time during the day to worry, rather than worrying about things all day long. Stroking a pet helps reduce stress, so does exercise (which reduces cortisol levels), and making sure you get enough sleep.

The trouble with procrastinating is that it results in more work that needs to be done next time you're in the office. And you have less time to do it, which adds to your stress, which makes you more likely to procrastinate.

One way to get started on your workload and not procrastinate even more is 'low hanging fruit'. It means you can start work on easier tasks that don't take too long. That results in you getting work done that isn't too hard, and it will help reduce your stress level. And once you've started work, it's easier to carry on with the larger and more difficult tasks.

What else can you do to deal with procrastination? Firstly, you give yourself a reward. That might be looking at social media or eating chocolate, but rewards help build new habits – like doing your homework. Secondly, spend time with people who get on with their work, rather than people who are prone to do anything else rather than work. That gives you good role models and they are likely to positively reinforce your better behaviour. What you don't need (although everyone tells you that you do!) is more willpower. Willpower is a very limited resource, apparently, so that's why it's better to get into good habits rather than relying on it time and time again because it runs out quickly.

So if it's you that's procrastinating, here are some good ideas. If it's a client that's having problems, you can perhaps suggest some of the ideas in this chapter to them.

References:

https://www.bakadesuyo.com/2012/05/12-ways-to-eliminate-stress/

https://www.thriveglobal.com/stories/29693-neuroscience-discovers-5-things-that-will-make-you-happy

https://www.bakadesuyo.com/2015/01/how-to-stop-procrastinating/

Stress in the workplace

Some facts about stress at work.

For many of our clients, work can be an unpleasant experience because of stress and that can lead to even more mental and emotional problems. It's said that if an employee goes off with stress, unless they return to work within six weeks, for every week after that there's a 10 percent less chance of them ever returning. Let's have a look at some of the UK figures for mental health issues.

According to the *Mental Health at Work Report 2017*, three out every five employees (60 percent) have experienced mental health issues in the past year because of work. Almost a third (31 percent) of the workforce has been formally diagnosed with a mental health issue. (The figure is up slightly from 29 percent in 2016.) The most common diagnosis was depression or general anxiety. 6 percent of employees have been living with a formally-diagnosed condition for over 10 years.

On the plus side, 53 percent of employees feel comfortable talking about mental health issues like depression and anxiety at work, and 76 percent of those who have experienced a mental health issue as a result of work feel that colleagues care about their wellbeing. On the down side, just 13 percent felt able to disclose a mental health issue to their line manager. Those who do open up, put themselves at risk of serious repercussions. Of those employees who disclosed a mental health issue, 15 percent were subject to disciplinary procedures, demotion, or dismissal (nearly double the 9 percent in 2016).

While 84 percent of managers accept that employee wellbeing is their responsibility, less than a quarter (24 percent) have received any training in mental health, although 49 percent of managers would welcome some specific basic training in mental health. From the employee's perspective, only 58 percent feel that their line manager is genuinely concerned about their wellbeing.

Separately, it's suggested that employees are reluctant to disclose mental health issues for fear of:

- Appearing weak
- Not being promoted
- Being selected for redundancy.

According to the Centre for Mental Health, the economic and social costs of mental health problems are high. From work they did in 2010, they suggest that the total cost of mental health problems in England in 2009/10 was £105.2 billion. NHS England spent £11.6 billion on mental health in 2016/17.

Stress can lead to suicide. In 2014, according to the Office for National Statistics, male suicide accounted for the bulk of all suicides in the UK, with rates more than three times higher than female suicide rates, at 18.2 deaths compared with 5.2 female deaths per 100,000 population. The highest rate of suicide was seen in men 40 to 44 years of age. Approximately 90 percent of young people who die by suicide have a mental illness, with more than half having major depression.

Mental health and mental illness are different things. People diagnosed with a mental illness can still have high levels of general mental wellbeing, while others without a diagnosed mental illness can show low levels of mental well-being. It can be useful to think of mental health as a matrix, where people can move among states of mental well-being regardless of any mental illness. The mental health continuum is shown in Figure 1.

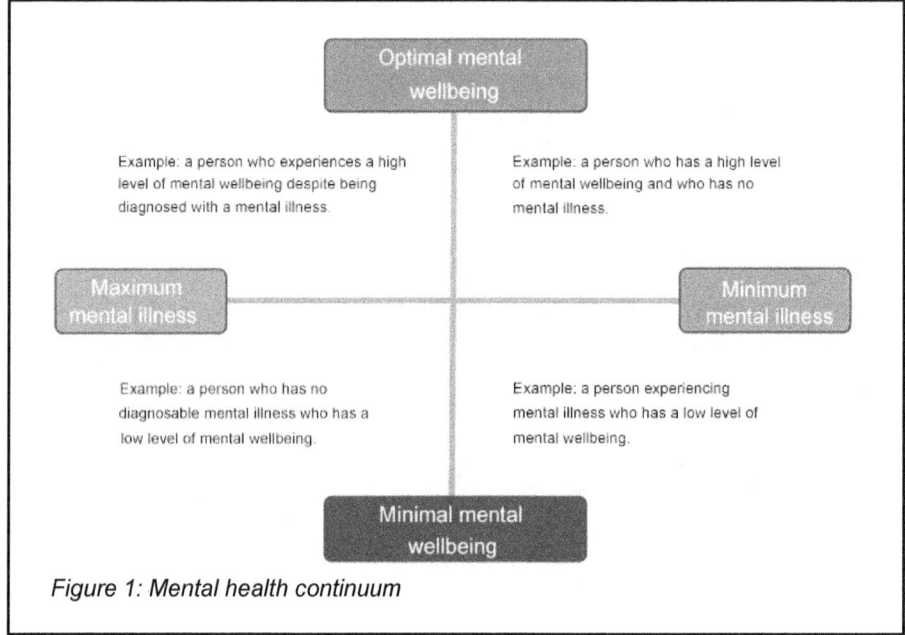

Figure 1: Mental health continuum

The interesting thing is that people don't live in any quadrant all their life and may find, for example, that they go from top right to bottom left for a period of time, and then back again. However, some people in that bottom left quadrant can find themselves staying there for long periods of time. One reason given for people finding it hard to change is stigma.

It's estimated that 78 percent of mental health problems started when the person was in their teens. And in 2016, it was estimated that 80 percent of students had mental health problems in the previous year.

People have a metaphorical stress bucket that can be filled up with stress. If more stress goes into the bucket than can be dealt with, it will overflow, and that leads to emotional problems developing. As well as an overfull bucket, the other issue that people have to contend with is that different people have differently-sized buckets. So what might be manageable stress for one person (with their big bucket) may prove too much for someone else (with their smaller bucket). Helpful coping strategies for emptying your bucket include:

• Sleeping (although stress often leads to insomnia, which then makes things worse).

- Physical activity.
- Walks in the countryside.
- Stroking a pet.
- Meditation, which can increase the size of the hippocampus – the part of the brain associated with memory.

So what can you do when someone comes to see you saying that they have had a mental health issue (and remember it's very hard for anyone – particularly men – to admit to any kind of problem)? The simple answer is to say that they are not alone, to show them that you are supportive. The secret is to treat people as you would want to be treated, and remember that people do recover.

For the organization that person works for, they need to recognize that there are business benefits from looking after the wellbeing of staff. If one member of staff goes off with a stress-related illness, then their work is shared by their colleagues – adding to those colleagues' stress. And that could lead to more staff going off with stress-related issues. The company could get some temporary staff in, but one of the more experienced staff will have to be released to train the new people. And while the training is going on, the work they would have been doing is shared among the few remaining staff – increasing the likelihood that they will become stressed. So, putting in place processes to prevent the first person going off with stress can save a company more money than they cost.

Nowadays, people are more likely to talk about stress, their feelings, and emotional issues. But there is still a feeling that people should be able to cope on their own and that leads some people to shun the company of others and spend a lot of time on their own, listening to their own thoughts, which probably aren't helping them make the best choices for getting help with their mental health issues.

While hypnotherapy may not be able to help cure mental health issues, it can help people with sleep issues, and it can help people empty their stress buckets and deal with the difficulties they are experiencing. Hypnotherapy can't be used with people whose condition means they have a weaker grasp on reality. But around your clinic and you friends and family, it can be worth keeping an eye out for people who are starting to show signs of a mental health problem.

References:

https://www.uk.mercer.com/our-thinking/health/mental-health-at-work-2017-report.html

https://www.centreformentalhealth.org.uk/economic-and-social-costs

https://www.thecalmzone.net/2014/02/onssuicidereport/

http://www.togethertolive.ca/mental-health-continuum

https://www.theguardian.com/higher-education-network/2016/mar/02/student-mental-health-a-new-model-for-universities

https://fullfact.org/health/nhs-spending-mental-health/

Dealing with fear of needles

A phobia of needles can't be dealt with in quite the same way as other phobias.

Let's start this discussion by looking at how people get phobias in the first place. There is some evidence that fear of spiders and fear of snakes might be innate, but how do people become scared of blood or potatoes or anything else? One way is through associative learning. This kind of learning is based on a stimulus that has a positive or negative consequence. And the more often the stimulus is presented, the greater the reinforcement of the resulting learned behaviour. There's also evidence that vasovagal needle phobia is a genetic maladaptive response that can't be unlearned. It's an inherited vasovagal reflex of shock, triggered by a needle puncture. People who inherit this reflex (yes, it seems to run in families) can grow to fear needles through successive needle exposure. What that means is that most phobias will disappear with a Rewind (fast phobia cure), but not vasovagal needle phobia.

> Fear of needles is sometimes referred to as aichmophobia or belonephobia, which really mean fear of sharply-pointed objects.

It is estimated that around 10% of the population have a fear of needles, although the figure is likely to be higher because sufferers tend to avoid medical treatment and so go uncounted. Sufferers choose to avoid inoculations, blood tests, and, sometimes, all medical care.

Types of needle phobia include the following:

Vasovagal

Around 50 percent of people with needle phobia have an inherited vasovagal reflex reaction. About 80 percent of those people have a close relative with the same disorder indicating that it's genetic.

These people fear the sight, thought, or feeling of needles or needle-like objects. This leads them to faint (vasovagal syncope) because of a drop in blood pressure. It's now thought that an initial episode of vasovagal syncope during a procedure with a needle is probably the primary cause of the needle phobia rather than any basic fear of needles. The condition starts with momentary high blood pressure and a fast heart rate (a fight or flight response) followed by both decreasing enormously at the moment of injection. In some cases, the drop in blood pressure caused by the vasovagal shock reflex may cause death.

CBT techniques of desensitizing and exposure have been used in the past. Nowadays, the 'applied tension technique' (developed by Lars-Göran Öst) is used to prevent fainting. It involves tensing muscles in the body, which then raises blood pressure. If a person's blood pressure increases, they are less likely to faint.

Associative

Associative fear of needles affects 30 percent of needle phobics. As mentioned above, a traumatic event causes the person to associate all procedures involving needles with the original negative experience. This type of fear of needles can lead to extreme unexplained anxiety, insomnia, preoccupation with the coming procedure, and panic attacks.

According to the NHS, you have an injection or needle phobia if:

- You have a marked, persistent, and excessive fear of needles.
- Exposure to needles almost invariably provokes in you an immediate anxiety response.
- You recognize this fear is excessive.
- Needle-sticks are either avoided, or endured with intense anxiety or distress.
- Avoidance, anxiety, or distress significantly interferes with your normal routine, occupational or academic functioning, social activities or relationships, or there is severe distress about having the phobia.

Resistive

Resistive fear of needles affects 20 percent of people affected and occurs when the underlying fear involves not simply needles or injections but also being controlled or restrained. It's said to be cause by a repressive upbringing or poor handling of prior needle procedures, perhaps with forced physical or emotional restraint. Symptoms include combativeness, high heart rate coupled with extremely high blood pressure, violent resistance, avoidance, and flight.

Hyperalgesic

Around 10 percent of needle phobics have a hyperalgesic fear of needles. That means these people have an inherited hypersensitivity to pain (hyperalgesia). So, the pain of an injection is unbearably great. The symptoms include extreme explained anxiety, and elevated blood pressure and heart rate at the immediate point of needle penetration or seconds before. Usually some form of anaesthetic helps these sufferers.

Just to make it more difficult, some people experience more than one kind of needle phobia.

So, to be clear, the term needle phobia is commonly used, even among medical professionals, to describe several very different conditions. For many people, the needle is only a source of fear because a needle is a necessary part of the procedure that provokes a frightening involuntary reaction in their body. Some sufferers of needle phobics have no fear of needles at all, but are extremely frightened of suffering the physical effects of a needle phobia reaction.

Needle phobia is unusual for a phobia in that it is a direct cause of death in many documented cases – and probably the cause in many many more undocumented cases because of all the people who avoid medical and dental treatment because of the condition.

When it comes to treatment, the usual relaxation session, rewind, and reframe will work on only about half the people suffering from needle phobia. And relaxing people who are likely to faint is probably not a good idea. But identifying which kind of needle phobic they are can help to determine the course of treatment. People with vasovagal needle phobia can be taught the applied tension technique to prevent them fainting.

References:

https://en.wikipedia.org/wiki/Fear_of_needles

https://makaylaheisler.wordpress.com/2013/08/05/how-is-a-phobia-formed-through-associative-learning/

https://www.anxietybc.com/parenting/tools/applied-tension-technique

http://www.needlephobia.com/

https://www.cnwl.nhs.uk/wp-content/uploads/Needle-Phobia-Booklet.pdf

Helping people with addiction

What is addiction and what can be done about it?

People can become addicted to pretty much anything that once seemed like fun to them. People can become addicted to drugs, alcohol, gambling, work, shopping, sex – in fact any number of things. And it's been suggested by the charity Action on Addiction that 1 in 3 people are addicted to something.

Addiction is defined as not having control over doing, taking, or using something to the point where it could be harmful to a person. Managing an addiction can seriously damage a person's work life and relationships. In some cases (with drugs and alcohol for example) the addiction can result in serious psychological and physical effects. For some people, an addiction can be a way of blocking out difficult issues.

Addictive stimuli are:

- Reinforcing – they increase the likelihood that a person will seek repeated exposure to them.
- Intrinsically rewarding – they are seen as being inherently positive, desirable, and pleasurable.

People are likely to become addicts because of genetic or environmental risks. Studies estimate that genetic factors account for 40–60% of the risk factors for alcoholism. Studies performed on twins found that rarely did only one twin have an addiction. In most cases, where at least one twin suffered from an addiction, both did, and often to the same substance. Environmental factors include:

- Age – adolescence is a period of unique vulnerability for developing addiction.
- Comorbid disorders – people with co-occurring mental health disorders (eg depression, anxiety, ADHD, or PTSD) are more likely to develop substance use disorders.

People don't start out as addicts, there are stages. These are:

- Experimentation – there's the curious stage when people are simply trying something new.
- Regular use – people actively try to recreate the experience. At this stage, quality of life is not affected.
- Increased use – an occasional behaviour escalates to frequent use. Risky behaviour may occur.
- Dependence – the person can no longer function normally or happily without taking a particular substance or

> There's a difference between drug addiction and dependence. Drug dependence is a disorder where stopping taking a drug results in an unpleasant state of withdrawal, which can then lead to further drug use. Addiction is the compulsive use of a substance or performance of a behaviour that is independent of withdrawal.

carrying out an activity. Withdrawal symptoms are strong and, despite negative consequences, the user cannot stop.

Because addiction affects a person's brain's executive functions, people with an addiction may not be aware that their behaviour is causing problems for themselves and others. Over time, craving the pleasurable effects of a substance or behaviour may dominate all of a person's activities resulting in them losing control over its use. The good news is that research suggests recovery is the rule rather than the exception. For example, over 15% of soldiers in the Vietnam war were addicted to heroin by the time they got back to the USA after the war. More than 95% of those soldiers stopped using heroin once back in the USA.

In normal circumstances, the neurotransmitter dopamine is what conditions us to do the things we need to do (like eat etc). An increase in dopamine levels motivates us to perform these actions.

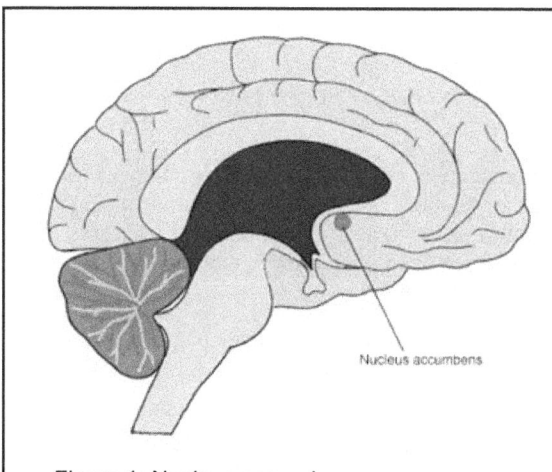

Nucleus accumbens

Figure 1: Nucleus accumbens

So what's actually going on inside an addict's brain? Anything pleasurable releases dopamine in the nucleus accumbens, which is referred to as the brain's pleasure centre. All addictive drugs cause a surge of dopamine in the nucleus accumbens. Addictive drugs raise the dopamine level by as much as five or 10 times the normal amount.

The probability that a drug or behaviour will lead to addiction is directly linked to how quickly dopamine is released, the intensity of that release, and the reliability of the release. When the nucleus accumbens gets extra dopamine, the hippocampus remembers the rapid sense of satisfaction, and the amygdala creates a conditioned response to the stimulus. And that's where things change from liking something to becoming addicted to it.

It seems that dopamine interacts with glutamate (another neurotransmitter) to take over the brain's system of reward-related learning. Repeated exposure to an addictive substance or behaviour causes nerve cells in the nucleus accumbens and the prefrontal cortex that couples liking something with wanting it,

Chasing the dragon often refers to the fact that subsequent hits of opium are never as good as the first one, so the addict craves more and more to get the same effect.

Tolerance – the diminishing effect of a drug resulting from repeated administration at a given dose.

in turn driving us to go after it. It means that people become motivated to seek out the source of pleasure.

Dopamine itself doesn't makes you feel good, the feelings of pleasure comes from opioids in the brain, neurochemicals which increase pleasure and deaden pain. Dopamine helps your brain to recognize what's called 'incentive salience' (a type of motivation resulting from an association between a certain stimuli and reward). And that means a person will want to do it again, and again – it becomes a conditioned reflex.

Over time, the consistently high levels of dopamine results in desensitized neurons (so that they are less affected by it), and fewer receptors. So a person's brain requires more dopamine than it can naturally produce, and it becomes dependent on the drug, which never actually satisfies the need it has created.

What can you do for people who want to stop being addicted? The answer is all the usual things of helping them relax and reduce the stress in their life, helping them to empty their stress bucket, helping them to sleep. In addition, scaling can help them to see they are making progress, and the miracle question can help them to visualize a life without the addiction.

Good questions to ask them include:

- What's changed since you first contacted me and now?
- For you, what's the most important reason to change your behaviour?
- When do you plan to stop?
- What do you plan to do instead?
- Currently, when don't you need to take the drug/perform the behaviour?
- Why isn't your addiction worse?
- What strengths have helped you?
- How have you coped?

If an addict is seeing you, make sure that they want to change and aren't being sent by someone else or just seeing you to complain about the NHS or their CBT sessions. Assuming they do want to change, then solution-focused hypnotherapy is the best method to help them.

References:

https://www.nhs.uk/live-well/healthy-body/addiction-what-is-it/

https://www.psychologytoday.com/gb/basics/addiction

https://www.psychologytoday.com/gb/blog/women-who-stray/201701/no-dopamine-is-not-addictive

https://www.helpguide.org/harvard/how-addiction-hijacks-the-brain.htm

http://bigthink.com/going-mental/your-brain-on-drugs-dopamine-and-addiction

http://coping.us/images/Solution_Focused_Therapy_for_Alcohol_and_Substance_
Use_Disorder_2-19-16.pdf

https://en.wikipedia.org/wiki/Addiction

https://www.hypnotherapy-directory.org.uk/articles/addictions.html

Useful NLP techniques

This 'vademecum' contains a number of NLP techniques that can be used by solution-focused hypnotherapists.

Before I start/one more thing

These are simple techniques to use. Your client has come in and sat down and you've chatted about the weather/the traffic/parking and you say the magic words; "before we start". Your client is still relaxed. They haven't put up the protective barrier that they intend to use with you to keep themselves safe. And you can then start your session – their drawbridge is down and their portcullis is up and you can march straight into their castle! You can do powerful work.

Similarly, you can suggest that the session is over (in the style of TV detective Columbo) and then say: "just one more thing". Suggesting that the session has finished means that they will breathe a sigh of relief and relax. Again, their defences are down, and you can do more work with them, or find that their answers reveal more about what's going on internally.

Building rapport

Rapport is a process, not a thing. It is something we do with another person. There are three elements to rapport: mutual attention (where each person is tuning in to the other); shared positive feeling (mostly conveyed by non-verbal messages); and synchrony (people unconsciously respond to each other's movements and gestures).

People like people who are like them. There are various techniques to achieve this effect.

Commonality is the technique of deliberately finding something in common with a person in order to build a sense of camaraderie and trust. This is done through shared interests, dislikes, and situations.

Mirroring and matching is another way of rapport building. Mirroring is the simultaneous 'copying' of the behaviour of another person, as if reflecting their movements back to them in a mirror. Matching can have a built-in 'time lag'. So, for example, if a seated client

Jargon buster

Submodalities – subsets of the modalities and the building blocks of the representational systems by which we code, order, and give meaning to the experiences we have.

VAKOG – Visual, Auditory, Kinesthetic, Olfactory, and Gustatory modalities

Resourceful state – what they want (confident)

Unresourceful state – how they are (nervous)

Associated – emotionally involved

Disassociated – like watching someone else.

uncrosses their legs and leans slightly inward while speaking, you can wait for a few seconds and then discreetly adopt the same posture.

Things you can match with mirroring include:

- Body posture – matching the angle of the spine works well and is not obvious.
- Breathing – including the rhythm, how deep or shallow, and whether from the chest or the abdomen.
- Voice tone – for example volume, speed, tonality, and speech rhythms, but not accent.
- Crossover matching – matching someone's gestures with a different part of your body. People do things like scratching their chin, flicking their hair, crossing their legs, and you can match this subtly by some equally natural-looking movement like tapping a pencil or jiggling your foot.
- Representational systems (VAKOG)
- Metaphors – use the same type as the client (eg gardening, fishing, sport, etc).

Directly matching gestures can be counter-productive because people can spot it very easily.

Emotional mirroring can also be used to create rapport. You empathize with someone's emotional state by being on 'their side'. It's important to listen for key words and problems that arise, so you can talk about these issues and question them to better your understanding of what they are saying and show your empathy towards them (Arnold, E and Boggs, Josh. 2007).

Once you have built rapport with someone you are in a good position to influence them and can make your suggestion in a climate of rapport where it will be greeted favourably.

Pacing and leading

Pacing and leading someone is much the same as the matching technique. So, matching someone's breathing rate is the same thing as pacing their breathing. After pacing someone's subconscious behaviours for a while you can begin to lead them.

Another way to pace someone is to affirm things you know to be true about that person's model of the world or their thoughts on what is true and verifiable. Here are some pacing and leading examples:

- As you continue to breathe … you can begin to learn the art of persuasion
- As you sit in your chair … you may realize a new way to use this technique
- As you feel your clothes against your skin … you will feel more confident.
- As you listen to my voice … a sense of understand will begin to develop.
- As you feel the need to blink … you may think of new ways practice persuasion.
- As you look at me … you may feel a deep connection.
- You can see these words … as you think of something wonderful.

- You can smell breakfast … as you begin to hunger for more persuasion techniques.
- You can see the sun … as you focus on your inner thoughts.
- You can hear my voice … and feel the power of persuasion.
- You may blink … before you realize how much you have learned.
- You may be sitting … but your mind can begin to wonder, what's next?
- The time is now … when you realize how much you have learned.
- It is time now … to thank you subconscious mind for all it does for you.

Pacing statements are very powerful in leading a person's emotional response and any statement that can be verified is a pacing statement. Pacing and leading is one of the fastest ways of eliciting emotion. After you have paced someone you can lead them. Lead them with your voice and language as well as with your body language.

Another way to pace someone is by repeating their key words or phrases. These are the words or phrases that person uses repetitively. If you notice that someone says "that's right" after almost everything you say then pace their language by expanding your vocabulary to contain their repetitive phrase. Then periodically and tactfully reflect their language back to them. "That's right", you do understand how to pace and lead people to new and exciting feelings.

If you want to give your leading statement even more influence, stack your pacing statements. Name two things you know to be true then add one you want to be true. As you read this and notice the movement of your eyes, you take a breath, and realize that you knew how to be persuasive all along. This is called a 'yes set' and is a common sales technique. If a salesmen can get you into the habit of saying "yes" then he can get you to buy a product by saying yes on cue by adding the question to the end of a 'yes set'.

Peripheral vision relaxation

This is a great way of relaxing in situations that are becoming stressful. Usually, we use 'foveal' vision, where we concentrate on one point in front of us and notice all the details about that one point, and ignore everything around it. It's an area of the eye with photoreceptors that are mainly cones (which see colour). 'Peripheral' vision, takes in everything that's happening in front *and* around us. It uses different light receptors in the retina and different neural pathways in the brain. Outside the foveal area (except for the blindspot), the photoreceptors are mainly rods (which are used to see in dim light). Foveal vision is linked to arousal of the sympathetic nervous system (fight and flight)) while peripheral vision is linked to parasympathetic arousal (rest and digest).

Here's what you do:

Just find a point straight in front of you and up a bit, and focus on it. Now gradually become aware of what's around it... and let your vision spread out in front of you to the corners of the room, as your eyes continue to look at that point and you become more and more aware of the periphery of your vision. If you stretch out a hand to one side of you, you might find the point on the edge of your vision where you see only that hand when you wiggle the fingers. Let your awareness also spread behind you …

just let your senses of hearing, touch, smell, and spatial awareness spread out to the periphery as well ... and notice how you feel changes ...

With peripheral vision, your client should experience certain physiological changes, a change in their breathing from higher to lower in the chest, their face and jaw muscles relax, and sometimes their hands became warm. If they usually have an internal dialogue going on, they might notice it becoming quieter or stopping altogether.

Circle of excellence

The Circle of Excellence technique was originally developed by Dr John Grinder as a basic self-anchoring process that can be used to elicit, create, and stabilize desired states. It allows a person to access the resource or skill they need just when they need it.

Here's how to use it:

1 Imagine a circle on the floor.
2 Think of a future challenging situation for which you want this resource.
3 Imagine yourself there; what do you see, hear, and feel? Who else is there and what are they doing and saying. Make the situation as clear and detailed as you can.
4 Think about the skills you need – confidence, calmness, self-control, etc.
5 Recall past experiences (or imagine) when you really excelled at that skill.
6 Now step into the circle and relive that past experience. Involve all your senses; see yourself doing it, saying what you said, feeling what you felt.
7 Now see yourself in the future using these skills for the challenging situation you thought of at the beginning.
8 Step out of the circle and think whether there are any other resources you need. If there are, you can repeat the process until you have all the skills you will need.

Later, when you are in the challenging situation, you can imagine you are standing in that circle of excellence drawing on all those skills. You can have a different circle of excellence for different situations.

The Swish technique

This technique allows you to instantly replace a recurring negative thought or behaviour with a positive one. Here's what to say with your client either sitting talking to you or nicely relaxed on the couch:

And while you're so completely relaxed, I'm going to teach you the Swish technique. Now I want you to imagine that in front of you is a huge cinema screen. And we're going to see still pictures on that screen in a moment.

Right, thinking about that interview next week (or whatever) ... I want you to create two pictures – which we're going to project on that cinema screen in front of you. They are going to be full colour, brilliantly lit pictures filling the whole of that screen.

The first picture is called 'The Moment of Anxiety', and is of the scene as you sit in the waiting room, waiting to be called into the interview – probably the moment when you're most anxious. Make that picture as detailed as you can: the room, the furnishings, the other candidates, their expressions, the scenery, sounds, smell, touch, etc – everything. When you've got that picture vivid enough that it actually makes you squirm, then you've got it right. It should be like a photograph taken through your eyes, full colour and brilliantly lit. When you're looking at that detailed picture on the screen, just move your hand for me.

<moves hand>

Good, now put that picture to one side for a moment, we shall be coming back to it in a minute.

Now for something much more comfortable. This time you are going to create an image of you just at the moment when you have SUCCESSFULLY been offered the job. Again make it as vivid as is humanly possible, and do the same 'tricks' as before to make it truly lifelike, with the interview panel wanting to shake your hand, the headteacher's office, and so on. We will call this one the 'The Moment of Achievement'. In it, you will be looking truly successful. Picture the look on your face. The interview has gone well, you actually enjoyed it, you did your best, and you feel really good. And there it is written all over your face. Allow yourself to enjoy it for a moment … then imagine the picture shrinking, becoming smaller and smaller, with the colours becoming less-and-less pronounced, until you are left with a small black-and-white picture the size of a postage stamp. Then 'lay it to one side', just as you did the first one. When you have done that just move your hand again.

<moves hand>

Excellent. Now I want you to have both of those pictures on the big cinema screen in the following way. The first picture, 'The Moment of Anxiety', in full colour and brilliantly lit filling the whole of that cinema screen except for one corner, where, tucked in, like a snapshot tucked into the frame of a larger picture is a small, dull black-and-white picture, picture number two, 'The Moment of Achievement'. Just arrange those pictures in that way. When you are ready just move your hand.

<moves hand>

Excellent. Now <speed it up!>, in a moment I shall say the word "Swish", and when I do, I want you to Swish those pictures over. The small becoming large, the large becoming small, the small becoming full colour, the large becoming black-and-white, the small becoming brilliantly lit, the large becoming small, dull and black-and-white. So, when I say the word "Swish" just Swish them over.

OK, Swish!

Now allow yourself to enjoy the feelings that are displayed on your face in that large picture for a couple of moments.

Brilliant – well done. OK, let's start again. Let's have those pictures on the screen. The first picture, 'The Moment of Anxiety', in full colour and brilliantly lit filling the whole of the screen except for one corner, where, tucked in, like a snapshot tucked into the frame of a larger picture, is a small dull black-and-white picture – picture number two, 'The Moment of Achievement'. Just arrange those pictures in that way again. When you are ready just move your hand.

<moves hand>

Good and Swish. The small becoming large, the large becoming small, the small becoming full colour, the large becoming black-and-white, the small becoming brilliantly lit, the large becoming small, dull, and black-and-white.

So whenever I say Swish just Swish them over, and allow yourself to enjoy the feelings that are then displayed on your face in the large picture for a couple of moments.

Let's start again. The first picture, 'The Moment of Anxiety', in full colour and brilliantly lit filling the whole of the screen except for one corner, where, tucked in, like a snapshot tucked into the frame of a larger picture, is a small dull black-and-white picture – picture number two, 'The Moment of Achievement'. Just arrange those pictures in that way again. When you are ready just move your hand.

<moves hand>

And Swish. The small becoming large, the large becoming small, the small becoming full colour, the large becoming black-and-white, the small becoming brilliantly lit, the large becoming small, dull, and black-and-white.

Let's start again. The first picture, 'The Moment of Anxiety', in full colour and brilliantly lit filling the whole of the screen except for one corner, where, tucked in, like a snapshot tucked into the frame of a larger picture, is a small dull black-and-white picture – picture number two, 'The Moment of Achievement'. Just arrange those pictures in that way again. When you are ready just move your hand.

<moves hand>

And Swish. The small becoming large, the large becoming small, the small becoming full colour, the large becoming black-and-white, the small becoming brilliantly lit, the large becoming small, dull, and black-and-white.

Let's start again. The first picture, 'The Moment of Anxiety', in full colour and brilliantly lit filling the whole of the screen except for one corner, where, tucked in, like a snapshot tucked into the frame of a larger picture, is a small dull black-and-white picture – picture number two, 'The Moment of Achievement'. Just arrange those pictures in that way again. When you are ready just move your hand.

<moves hand>

And Swish. The small becoming large, the large becoming small, the small becoming full colour, the large becoming black-and-white, the small becoming brilliantly lit, the large becoming small, dull, and black-and-white.

Let's start again. The first picture, 'The Moment of Anxiety', in full colour and brilliantly lit filling the whole of the screen except for one corner, where, tucked in is a small dull black-and-white picture – picture number two, 'The Moment of Achievement'. Just arrange those pictures in that way again. When you are ready just move your hand.

<moves hand>

And Swish. The small becoming large, the large becoming small, the small becoming full colour, the large becoming black-and-white, the small becoming brilliantly lit, the large becoming small, dull, and black-and-white.

OK, now I want you to do the same thing yourself another five times. You say the word Swish in your head and Swish that moment of anxiety picture and replace it with the moment of achievement picture. And each time just enjoy the look on your face when you're offered the job. And when you've Swished five times in your head, just move your hand to let me know.

<hand movement after a minute>

Excellent – well done. That's the Swish technique. You'll be able to use it again whenever you want to, in the future.

And now … *the rest of the session.*

A nice alternative for emergencies is the one-handed Swish:

1 Have the unwanted scene (Moment of Anxiety) on the back of your hand. Describe to yourself what you are seeing.

2 Turn your hand over and have the wanted scene (Moment of Achievement) on the inside of your hand (palm). Adjust the scene to intensify it.

3 Start with the palm of your hand facing away from your face, about six inches away. Very slowly move your hand as far as your arm will extend.

4 As slowly as possible, turn your hand so that the palm side is facing your face, keeping your arm extended.

5 Bring your palm up to your face.

6 Repeat this procedure five times.

Here are some tips to get it to work as well as possible:

• The repeated Swishes can get faster and faster each time.

• Between each Swish, think of something completely different or open your eyes. This will break state, so that the process goes only one way. This creates a one-way 'chain', a direction pattern as you change size, brightness, and association/dissociation of your image.

• When you Swish the images, it is often helpful to make a "Swish" sound or say "Swish!" out loud.

• The final test (or 'future pacing' in NLP-speak). Conjure up 'The Moment of Anxiety' image again. It should be hard to see. Instead, 'The Moment of Achievement' image has replaced it and therefore has replaced the behaviour/anxiety/etc.

Fast phobia cure (Rewind)

A way to turn a traumatic memory into an ordinary one. NLP fast phobia cure.

You can tell clients:

- It's a way for you to convert phobia memories to narrative (ordinary) memories.
- The memories move from the amygdala (in the primitive brain) to the cerebral cortex (intellectual brain). So, the phobia will be gone – completely.

It's an interesting and very effective technique. It involves you in the present looking at yourself in the past. You're going to be an observer of events that happened that upset you in the past. But in the present you won't find them upsetting.

What usually happens, once we've used this Rewind technique, is that you'll feel almost exhilarated by the fact that you're not feeling traumatized by your fear of snakes (or whatever) anymore. You'll feel free, no longer held back in any way by your phobia, and ready to move on with your lives

We need to use two memories – the first and the worst time you had the phobia. Or any two that are significant for you.

It's important that you ask yourself, what exactly are the keypoints in those memories. And here's the clever part, even if you can't remember all the details, you can let your imagination fill in the details. You need enough detail to create in your mind a short movie.

And you need to have a safe point at the start of the memory, and a safe point at the end of the memory.

What you're going to do in a minute is imagine that you're watching these memories on a big cinema screen – it's as if you've gone to the cinema and the film they're showing is a couple of events from your life! But then you're going to keep rewinding the film and watch the same events again and again. The good news is that by the end of the session, those memories will seem dull and boring and nothing to be frightened about at all – goodbye fear of snakes!

One other thing I'll need you to do, when you're on the couch, is signal when you're ready – or when you've completed a task. Basically, I need you to move your hand – a little wave – like this. Is that OK?

While we're doing this, I need you to be aware of what you're doing, but relaxed. So I will do a short relaxation-style introduction, and then you'll just imagine you're watching a movie.

OK, let's get you on the couch. *Note: the script uses a fear of snakes as an example.*

And I want you to start by thinking about the top of your head ... it's a good place to start ... it's a good convenient place to start ... it's also a place where tension often starts ... tension often starts in the little muscles of the scalp, so I want you to think about those little muscles and the skin of your scalp ... thinking about those little muscles in the skin of your scalp ... just allow them to let go and relax ... now think about all the muscles of your face, just let them go slack ... your forehead and your eyes and eyelids ... the cheeks, mouth, and jaw muscles ... it's a wonderful feeling when you let your face totally relax, because you can actually feel the skin settling, smoothing out ... it might mean that your mouth opens slightly, but whatever's best for you, just let it happen ... unclenching your teeth and relaxing your tongue, because the more you physically relax, the more you can mentally relax ... and the more you mentally relax, the more you physically relax ... thinking about your neck and shoulder muscles now, and into the tops of your arms, letting all tensions drain away ... I always imagine it's like an ocean of calmness ... an ocean of calmness just moving down through your body ... leaving you at ease and just as comfortable as you've decided you'd like to be ... that ocean of calmness ... moving down into your neck and shoulder muscles ... and the tops of your arms ... as you think on down through your elbows ... into your forearms ... down through your wrists and into your hands ... right the way down into the very tips of your fingers and tips of your thumbs ... just letting all those muscles let go and relax ... that ocean of calmness just letting you be as comfortable as you've decided you'd like to be ... and now think about your breathing, noticing that you're breathing even more steadily, even more slowly, as you relax more and more, so you can let any tension in the chest area simply drain away, as you think on down to your stomach muscles, letting those muscles relax, too ... think down into your back now, the long muscles either side of the spine, just let those muscles relax ... and your waist ... and your main thigh muscles, as you think on down through your knees, down through the shins and calves, just allowing all those areas to relax and let go, as you think on down through your ankles, through your feet, into the very tips of your toes ... that ocean of calmness leaving you at ease and comfortable ...

And I wonder whether you can picture yourself at the top of a short flight of stairs ... just five stairs going down ... and at the bottom of the stairs you can see the most comfortable chair you've ever seen, (with a remote control resting on the arm,) and near it is the best and biggest cinema screen in the world. You step on the first stair, you're allowing yourself to really relax and let go ... step 2 you feel yourself relaxing even more ... step 3 you're breathing is becoming so restful ... step 4 ... you can see that comfortable chair ... just one step to go ... Step five, you are so completely relaxed and you just snuggle into that comfortable chair (and take hold of the remote control) ...

The giant cinema screen is currently blank, (it's waiting for you to press the buttons on that remote control to play one of your memory tapes.)

(For very scared people.) Up above and behind you is a little projection room. The projection room is where someone changes the film reels and projects the image onto the screen. Now I want you to float out of your body up to the projection room in the cinema where you can watch yourself watching the screen. From this position you will be able to see the whole cinema including your head and shoulders sitting in the middle of the cinema. You can also imagine a clear Plexiglas barrier in front of you, which lets in all the sights and sounds. Put your hands on it and feel it while you watch what happens next in the memory movie. The film will play from beginning to end and tell the whole story of the memory (and you might imagine that because this is an old memory, what you see is in vintage fuzzy black and white).

You'll be able to watch that younger self on screen going through the experience. You can watch the movie as a detached observer, even as a stranger. You are completely safe, here and now, safe in the booth, feeling the glass with your hands, while also noticing the other self in the cinema seat watching. It's just a movie.

We'll play through the first memory-movie starting from a safe point at the beginning to a safe point at the end in a moment. This is your dream memory of being scared of snakes.

So now, when you can see that safe image at the start of the movie, just wave your hand. OK, remember we're at a safe point, the remote control is set to pause.

Let me explain what to do, and then I'll tell you to play the movie. This first time, I want you *(from the projection box)* to play through the movie in great detail. It's got to be bright and vivid, think about how things smell, think about the sounds going on in the background, think about anything you touch, how they feel, and, of course, think about your own feelings as you experienced the events in the movie.

If you feel too frightened or distressed, you can stop the movie instantly. Let me know if that happens by waving your other hand. If that happens, we'll go back to the beginning and try to play it a bit longer next time. Keep doing that until you can see yourself watching the whole movie all the way through.

When you've played through the movie and got to your safe place at the end, I want you to just wave your hand.

So that's the plan.

And now, I want you *(the projectionist)* to press play, and play through the movie in all its glorious detail.

This is a bad memory, allow the movie to be as bad as it was when you really experienced it. If we showed it to the rest of the world, it would be obvious just how distressing it was for you.

(If they finish too quickly, make them do it again, highlighting sensory information.)

Well done. Now this time (using your remote control), you're going to rewind the movie back to the beginning. You can still see the images, but now they look strange doing things backwards, and the sound is weird too. Wave your hand when you're back at the beginning.

Great. Now this time, fast-forward all the way to the end, pause for a while, and then rewind it again. You're still seeing all the images, just faster, and the sounds are all strange. Wave your hand when you're back at the beginning.

This time you're going to fast forward and rewind at an even faster speed. You'll notice how very strange it all seems. When you've done it, wave your hand.

Well done. OK, this time I want you to super fast forward and super fast rewind five times, then wave your hand when you've finished. You're going to need to concentrate.

You're doing really well. So, the same again, five times forwards-and-backwards at super fast speed. Wave your hand when you've finished. I don't know whether you've noticed how relaxed your body is now. You might even be getting a little bored.

Brilliant, well done. Now you're going to do the same again, five times super-fast forwards-and-backwards, but this time you can imagine the characters on the screen as cartoon characters perhaps, and you can imagine the soundtrack as being comedy music. Wave your hand when you've finished.

You've done really well. I'm making you work pretty hard, so well done you.

And we're half way through now.

So, you're still imagining that *(from the projection box you can see that)* you're sitting on a really comfortable chair, in front of a huge top-quality cinema screen, (with the remote control in your hand). And there's nothing on the screen at the moment ...

This time you're going to play through your second memory movie (going passed the snake shop) and we'll pause at a safe point at the beginning before watching it through to a safe point at the end.

So now, when you can see that safe image at the start of the movie, wave your hand.

Good. Again, I want you to play through the movie in great detail. It's got to be bright and vivid, think about how things smell, think about the sounds going on in the background, think about anything you touch, how they feel, and, of course, think about your own feelings as you experienced the events in the movie.

Then when you've played through the movie and got to your safe place at the end, I want you to wave your hand.

OK, I want you to play through the movie in all its glorious detail, now.

This is a bad memory, allow the video to be as bad as it was when you really experienced it. Again, if we showed it to the rest of the world, it would be obvious just how distressing it was for you.

(If they finish too quickly, make them do it again, highlighting sensory information.)

Well done. Now this time (using your remote control), you're going to rewind the movie back to the beginning. You can still see the images, but now they look strange doing things backwards, and the sound is weird too. Wave your hand when you're back at the beginning.

Good. Now this time, fast-forward all the way to the end, pause for a while, and then rewind it again. You're still seeing all the images, just faster, and the sounds are all strange. Wave your hand when you're back at the beginning.

Well done. This time you're going to fast-forward and rewind at an even faster speed. You'll notice how very strange it all seems. When you've done it, wave your hand.

OK, this time I want you to super fast forward and super fast rewind five times, then nod that you've finished. You're going to need to concentrate.

You're doing really well. So, the same again, five times forwards-and-backwards at super fast speed. Wave your hand when you've finished. Again, I don't know whether you've noticed how relaxed your body is now. And again, you might even be getting bored.

Terrific, well done. Now you're going to do the same again, five times super-fast forwards-and-backwards, but this time you can imagine the characters on the screen as cartoon characters, and you can imagine the soundtrack as being comedy music. Wave your hand when you've finished.

You've done really well. As I said, I've made you work pretty hard, so well done.

So, *(let's move back from the projection box to the real you)/(while you're)* sitting in that comfortable chair (with the remote control) looking at the cinema screen, let's imagine that you're watching yourself in a garden, perhaps the garden of a stately home, with friends and family, maybe sipping a cooling drink after walking round the grounds. Enjoy this movie for a while ... (or they could be on a beach).

(You could add a short relaxing trance session here if there's time.)

<pause>

And when you feel ready, you can come back to full awareness, feeling great, and able to open your eyes as soon as you're ready.

You did really well today, you took your fear of snakes and turned it into something that I think you felt was quite boring! You should feel very proud of your achievement. That was excellent.

The 5-minute headache cure

Firstly, remember that headaches and other pain could be an indication that something is wrong and a doctor should be consulted. With headaches, they could be caused by dehydration or injury.

Here's what to do:

1. Say to your client, "I know this is a strange question, but if this headache had a colour, what colour would it be?". Then ask, "if it had a shape (like a star or a ball, or whatever), what shape would it be?". Finally ask, "imagine the background has a colour that's different from the colour of the headache, what would that be?".
2. "Now gradually make the headache colour the same colour as the background so all you are left with is an outline of the shape. Visualize it happening now."
3. "Now make that shape smaller and smaller until it is just about to disappear."
4. "And now make it so small, that even if you try to look all over the background for it, you realize it has gone."
5. "Lastly, ask yourself, 'if there were a colour of healing, what would it be?' And then fill the background with that colour, top to bottom, right to left until it is completely full of that healing colour."

Anchoring

See *All you need to know about anchoring* in *Hypnofacts* page 61.

Cartesian Questions

Cartesian Questions can help your clients get new perspectives and understanding of what inspires and blocks them in making a decision or achieving a goal. The questions can help when considering all possible options around a decision or goal.

- What would happen if you did?
- What would happen if you didn't?
- What wouldn't happen if you did?
- What wouldn't happen if you didn't?

Dickens process

The Dickens process is a favourite of Tony Robbins and is based on Charles Dickens' book *A Christmas Carol*, where Scrooge is shown his potential futures.

This process uses your client's conscious mind to visualize two potential futures and attaches feelings to both alternatives, which is what makes the better choice likely to succeed.

Here's what you get your client to do:

Imagine they are at a fork in the road: one path is them continuing to do what they do (eg smoking) and the other path shows what will happen if they change a habit or remove a limiting belief (eg stop smoking).

You ask your client to imagine they choose the 'no change' path, and get them to imagine how bad things will be in 1 year, 5 years, 10 year, and 20 years. So, for a smoker it will be examining their health and perhaps their bank account. Get them to imagine not being able to play with their children/grandchildren because they are out of breath when they move quickly, etc. They must imagine the worst case scenario. Then, perhaps, imagine them having to break the news to their children that they're dying of lung cancer and they won't see them grow up. Get them to really feel what that would be like – experience the pain. The lack of happiness.

Now visualize going back to that fork in the road, and this time, choose the path of change – where a bad habit or limiting belief has been left behind like an old pair of shoes. They've stopped smoking. Get them to imagine how they'll feel in 1 year, 5 years, 10 years, and 20 years. They are much healthier, more successful because they now have more energy and feel stronger. They get to meet their grandchildren and play with them. They can see themselves enjoying life to the full. Get them to imagine the best possible outcome and the most positive version of themself. They should really see, hear, and feel it.

Then, come back to the present and that fork in the road and ask them which path in life they want to take. Because of how they felt at the end of each path, they are very likely to make the sensible choice.

The Godiva chocolate pattern

Also, see *Time for a change* in *Hypnofacts 5* page 48.

The Godiva Chocolate Pattern was developed by Richard Bandler, and transfers the strong motivation already held for one behaviour (eg eating chocolate) onto a less appealing behaviour, say going to the gym.

It's a simple technique that involves holding two pictures in your client's mind (one of the habit they are really feel motivated to do, and the other of the behaviour they are less keen on). Place the positive picture so it covers over the less well liked one. Then imagine a small hole in the picture gradually opening up until they can see the second picture clearly, all the time holding on to the feeling of being greatly motivated.

Here's how to do it:

1 Get your client to think of something they would like to feel more motivated to do.

2 What picture represents this activity for them? Get them to visualize this picture in their mind and call it picture 1.

3 Think of something they love to do (like eating chocolate), which they are highly motivated to do it. Call this picture 2.

4 Get them to notice all the details about it.

5 How does it make them feel? Make sure it makes them feel really motivated!

6 Now ask them to imagine a small hole is opening in picture 2.

7 As the hole in picture 2 gets bigger, they can see picture 1 (the picture of the activity they want to be motivated to do) behind it.

8 They should make that opening only as big as they can while maintaining the motivated feelings associated with picture 2.

9 As far as possible, make the details (submodalities) of picture 1 the same as picture 2 as that hole gets bigger.

10 When the hole is fully open, they will be looking at picture 1 (which previously they hadn't found very motivating) and feeling motivated!

Submodalities belief change

The steps are:

1 Identify a limiting belief. "As you think about that belief, what do you see?" (make them picture it and elicit the submodalities.)

2 "Think of a belief which is no longer true (eg Father Christmas, or 'I go to [name of junior school]'. What is it? As you think about that belief, what do you see?" (Elicit the submodalities.)

3 Change the submodalities of the limiting belief into the submodalities of a belief that is no longer true.
Test: now, what do you think about that limiting belief?

4 "Think of a belief which for you is absolutely true. Like, for example, the belief that you speak English (or whatever). Do you believe that? As you think about that belief, what do you see?" (Elicit the submodalities.)

5 "Think of a belief that you would like to have instead of the old limiting belief. What is it? As you think about that belief, what do you see?" (Elicit the submodalities.)

6 Change the submodalities of the desired belief into the submodalities of the belief that is absolutely true.

Test and future pace by asking, "Now, what do you believe? Why do you believe you have this new belief? What will happen next time you are in that situation where the limiting belief used to affect you?"

Key success factors in submodality work are:

1 Position yourself off to one side of the person you are working with, rather than directly in front of them, so that they have room to picture their representations and their eyes won't be fixed on you.

2 The better you calibrate the non-verbal responses of the client (eg where their eyes go when they picture their belief, changes in voice tone that show how certain they are about what they are saying) the easier and more successful the intervention will be.

Perceptual positions

This technique is used to change a client's perception or viewpoint about an experience in order to gather more information. It involves dissociating from the experience so they can get a different take on things.

Traditionally there were three positions, or steps, within the process, but Robert Dilts added a fourth, which some NLP coaches do not use.

You can use perceptual positions with a client for them to see the point of view of other people and to disassociate themself from emotions that may 'cloud' their perception of a situation. For example, there may be someone at work they don't get on with particularly well. They can do the perceptual positions exercise using that person in position 2.

There are four key steps, or positions, are:

1 Position 1 – through your own eyes
2 Position 2 – through the eyes of someone else
3 Position 3 – looking down on the whole scene (meta position)
4 Position 4 – meta to the meta position

Here's what you get your client to do:

In position 1, get them to imagine standing in front of the person they don't get on with. Be themself and tell that person exactly how they feel, what they're thinking, and what they think the problem is. When they have said absolutely everything they want to say, move them from where they were standing and break state.

For position 2, they need to stand in the place they imagined the other person was standing, and take on that person's physiology, voice tonality, and pretend to be them. This pretend version of that person has just heard everything your client said to them, so now it's their chance to respond. Your client should say whatever they think. When they've said everything they want to, you should move them from that position and break state.

In position 3, they should stand where they could have 'seen' the interaction that went on between positions 1 and 2. Your client should then say what they are thinking and feeling as if they'd just witnessed what both people said. Once they've finished, break state

Position 4 is like a fly-on-the-wall position. It's somewhere that could see all the other three positions without being involved. Here your client offers advice or recommendations to position 1.

Move your client back to position 1 and see how differently they feel.

Belief change using Sleight of Mouth

See *Arguing a point* in *Hypnofacts 3* page 26.

6-step reframe

Bandler and Grinder developed the six-step reframe technique from their study of Milton Erickson (ideomotor signals) and Virginia Satir's work with parts. They included it in their book *Frogs into Princes*.

Sometimes we adopt negative behaviours that have an underlying positive intention. For example, a client may drink too much for their health because they want to be seen to be sociable and they may think they are more relaxed and more fun to be with when they have had a few drinks. They want to stop drinking, but they don't want to stop going out with their friends and being sociable.

They need to be able to achieve their positive intention, socializing and having fun, without the negative side effect that is occurring. They need to find a positive behaviour to meet the same objective. Because the intention and the behaviour have become closely linked in their unconscious mind, they need to get the conscious and unconscious parts of the brain to communicate. And this can be done with the 6-step reframe.

Here's what to do:

1   Identify the behaviour they want to change. They need to be very specific so work out exactly what the behaviour is and get them to say it out aloud, acknowledging it. Remind them of the good aspects of the behaviour; what the behaviour has enabled them to do that was positive.

2   Now ask them whether they are ready to stop the behaviour and do something different. They need to communicate with their subconscious because that is the part that is choosing the behaviour.

3   Now separate the behaviour from the positive intention. Identify the positive intention behind the negative behaviour and clearly separate them in their mind so that one does not automatically link to the other but can be clearly separated.

4   Ask the client to think of other behaviours that would also satisfy the positive intention but that would not have negative side effects.

5   Get them to ask their unconscious mind whether it is prepared to change to the new behaviour.

6   Check that the new behaviour is ecological, that no-one is going to be put at risk or seriously inconvenienced by it.

Once they have used this process a few times they'll be able to use it to stop doing other things that they have grown out of!

Organic belief change technique

This technique is used to explore and reassess limiting beliefs. Organic Belief Change (aka Museum of Old Beliefs) is a technique using spatial anchoring to enable a client to question limiting beliefs that they hold.

Basically, a client identifies a limiting belief they would like to change, and a preferred belief. Check ecology, and then assist them to spatially anchor six positions. Some people include a seventh, a 'meta' or observer position. Break state in between each spatial anchor.

To implement the Belief Change Cycle, lay out on the floor separate cards for each of the states associated with the 'landscape' of belief change. This essentially involves having the person put themself fully as possible into the experience and physiology associated with each of these aspects of the natural cycle of belief change and 'anchoring' them to specific spatial locations. The six locations are:

1   'Want to believe' something new.

2   The experience of being 'open to believe' something new. If required, the client can identify a 'mentor' that helped them to become more 'open to believe' by 'resonating' with, releasing, or unveiling something deeply within them. Then make a physical space for the mentor near the 'open to believe' space. Mentors can include children, teachers, pets, people they've never met but have read about, phenomena in nature (such as the ocean, mountains, etc) and even themself.

3   The beliefs that they 'currently believe' now, including any limiting beliefs or beliefs that conflict with the new belief they would like to have more strongly.

4   The experience of being 'open to doubt' something they had believed for a long time. Again it's possible to identify another 'mentor' that helped the client to become more open to doubt something that was limiting them in their life.

5   Beliefs that they 'used to believe' but no longer believe. This is a space that is sometimes called the 'museum of personal history'.

6   An experience of deep 'trust'. This might be a time when they did not know what to believe anymore but were able to trust in themself.

Once this landscape has been laid out, it can be used in different ways. Usually, a client thinks of a new belief that they'd like to strengthen and simply steps through the cards, like this:

1   They stand in the 'want to believe' area and think of the new belief.

2   Keeping the belief in mind, they move to 'open to believe'. If they have a mentor for this state, they can step into their 'shoes'. Then seeing themself through their mentor's eyes, they can give the 'open to believe' version of themself helpful advice and support.

3   They stay there while they feel what it is like to become more open to believe this new belief until they feel ready, to move to 'currently believe', where they concentrate on the new belief they want to have.

4   At this stage, any conflicting or limiting beliefs may surface. They need to keep those beliefs in mind and move to 'open to doubt'. Here, if they have chosen a 'mentor', they can again step into their 'shoes', and, in the role of the mentor, give the 'open to doubt' version of themself helpful advice or support.

5   For an ecology check, they go to 'trust' and consider the positive intents and purpose of both the new belief and any conflicting or limiting beliefs. They consider whether there are any changes or revisions they would like to make to the new belief. And they should consider whether there are any parts of the old beliefs that would be worth retaining or incorporating along with the new belief.

6   Then they go back to 'open to doubt' (see stage 4) and revisit old limiting or conflicting beliefs, bringing the insights they had from 'trust'. They are then able to put the beliefs into 'used to believe' aka their 'museum of personal history'.

7   Then they go back to 'currently believe' (see step 3) and focus on the new beliefs they want to strengthen. They should try to experience their new sense of confidence and verbalize any new insights or learning that they may have discovered during this process.

8   As in step 5, they go to 'trust' for an ecology check, and consider the changes they have made.

People often find that by walking through these locations (or imagining walking through the locations) and re-experiencing the states allows them to completely shift their beliefs.

Doing an ecology check with a client is checking the consequences of their future actions and plans. There are basically four areas.

•   You should ask your client what will happen to them if they achieve their goal or belief? If some negative things will also happen (eg put on weight if they stop smoking) they will not be congruent, ie not all their thoughts, feelings, and beliefs will be in agreement and they won't do it.

•   You should ask how this change will affect others? Their goal may negatively impact their work or someone in their family.

- The next question is how will it affect society as a whole?
- Lastly, ask them what would happen to the planet if they achieved their goal/new belief?

Physiology of excellence

This is acting 'as if'. So rather than the outside world affecting our mood, we're behaving in a desired way until we find that's exactly how we feel – relaxed, confident, happy, etc.

Mend a broken heart

The procedure to use is as follows:

1 "Imagine you are sitting in the middle of a cinema and there is a still, colourful snapshot of you on the screen, as you look today, feeling comfortable, with a smile on your face."

2 Establish 3-place dissociation. "Now imagine that you float out of your body up to the projection booth where you can see yourself sitting in the cinema seat and also see the still, colour snapshot of yourself on the screen. You may even wish to imagine Plexiglas over the projection booth's opening, protecting you."

3 "Now, watch protected in the projection booth, as the other you in the cinema watches a black and white movie of a younger you going through the entire relationship – the good, the bad – from the first meeting to the end. Watch the whole event, starting before the beginning to the end. Observe until you are beyond the end of it, when everything was OK again.

   If you are not fully detached, make the cinema screen smaller and farther away. Now make the picture grainier and stop and start the film so that when you've finished watching it, you're completely detached. End the movie after the relationship event, with a freeze frame of yourself. Let me know when you are finished."

4 "Now bring back the still colourful snapshot of you feeling comfortable and smiling."

5 "Next I want you to leave the projection booth, float back down into the cinema, and step into the screen, into the very end of that movie and run that movie backwards in colour, full size, very quickly, in about 1.5 seconds, all the way back to before the relationship began. See, hear, and feel everything going backwards in those two seconds or less. Add some circus music, you may want to see your ex-partner with a clown nose, and feet. Do this two or three times."

6 "Now, with you still in the screen, run a full size, colour movie of all the bad things that happened in the relationship – the times that s/he let you down, made you angry, hurt your feelings, lied to you, etc – and let me know when you're finished."

7 "You can go back to your seat in the cinema and see on the screen that still, colourful photo of you, smiling and feeling good. And aren't you glad you're out of that relationship now? Aren't you glad s/he's living somewhere else?"

8 "Now I'd like you to think of yourself in the future, looking back on this relationship, realizing that you learned a lot from this experience. You can look back at it and

see that it turned out to be a great blessing in your life. And you have totally and completely forgiven [name of former spouse/girlfriend/boyfriend], and you have totally and completely forgiven yourself for any actions, words, or thoughts that were imperfect. Let me know when you are finished. Now think of yourself in the future, running into [name] somewhere, and tell me how it feels."

Allergy cure

Lots of us are allergic to what is in effect a harmless substance yet our immune system makes an inappropriate response to it. This technique lets the mind inform the immune system that we are no longer allergic.

Here are the steps:

1 "Imagine that your immune system has made a mistake and has mistaken something harmless for something that it needs to protect you from."

2 "Choose something similar to what you are allergic to. If you are allergic to cats but not dogs, then choose dogs for example."

3 "Now imagine you are in a room with the thing you are allergic to, in my example, a cat. Give a score out of 10 for how uncomfortable you feel."

4 "Remember though that your immune system has made a mistake and there is no need to have allergic symptoms in the presence of cats."

5 "Now imagine you are in a room with a dog. Give a score out of 10 for how uncomfortable you feel. You should feel relaxed and have no symptoms. Anchor that feeling by clapping your hands to reward your immune system for having the correct response this time."

6 "Imagine stroking the dog and playing with it, stopping every so often to clap to remind yourself that your immune system is behaving itself."

7 "Now imagine that a cat has just walked into the room and keep focusing on the dog and clapping."

8 "By now the clapping will be anchoring the non-allergic response, so use it when you are in contact with the object of your allergy."

The next thing to do is to test this out ,and remember to keep reinforcing the anchor whenever you can.

Parts integration/visual squash

Fear and uncertainty can hold back a client from fulfilling their potential. This may be because they had a bad experience in the past or they have a limiting belief about their ability. Your client ends up in the situation where part of them wants to make a change (eg lose weight) and part of them doesn't. This technique integrates the two parts.

Here's how:

1 "Identify the two parts that are in conflict and imagine they are people with hair, clothes and bodies, voices, and feelings."

2 "Ask the 'part' that is the problem to come and stand on one hand and the 'part' that is not the problem to stand on the other hand."

3 "Ask each part 'What is your positive intention and purpose?' So, for example, one part might say that it wants to save the other part from the fear of failure and wants it to have fun. The second part might say that being overweight is putting their health at risk and being ill wouldn't be fun."

4 "Allow each part to discuss how they can work together to a common positive intention."

5 "Once a positive intention has been agreed, ask each part what resources they can offer each other to reach this agreed positive intention."

This process allows a client to recognize that each part of their inner conflict has something to offer and that learning more about each part can bring your client to a desirable outcome that will integrate both parts. This will resolve their inner conflict and allow them to move forward in their decision with purpose and motivation, no longer held back by their inner self doubts.

Decision destroyer

This technique is used to change limiting decisions. Poor decisions can begin to ruin a client's life, where they are responding, in the present, to limiting programming of the unresourceful decision they made in the past. They are no longer functioning in their present reality.

Limiting decisions can go into a person's programming and become part of their mental map of who they are. They also control what your client will do in various situations. The problem is that even decisions that served them in past may have outgrown their usefulness and become outdated. This is the simple version of the Decision Destroyer pattern developed by Richard Bandler.

Here's what to do:

1 Identify an ongoing limiting decision. This could be a traumatic event.

   a Elicit it fully. When was it formed? How long have they lived with it? Fully associate to references underneath the belief.

   b What are the supporting events (references) that enhance the belief.

2 Go to a time that is before the event, at least 15 minutes before the limiting decision event (or trauma).

   a Stay in the previous event and ask them; "What experience could they have had then that would have prepared them for what was going to happen?"

   b And ask, "What were they deciding in this moment (before the event when the limiting decision was made)?

   c Go back before that at least 15 minutes...and before that, asking the question at each point, "What were they deciding in this moment?" until you get to a point in the past before the decision was going to be made.

d Notice the freedom they now have before the decision was even made. They should notice the possibilities that are now present in this moment before the limiting decision. Notice that they could have made a number of different decisions.

3 Come back through time evaluating all those experiences in the light of the new resources and new choices that they have made. Come all the way back to the present allowing the changes to occur until they are all the way back to the present having made those changes, fully and completely integrating them into their life.

4 Consider the present situation. Notice what choices have now become available and all the options they have now that they didn't have way back then.

5 As they consider their new choices. Ask them: "What is it in the future that you will feel now to allow you to be comfortable with your new decision(s) that you have made?"

6 Future pace. "Imagine yourself in the future in a specific moment when these new choices are operating and you have new behaviour operating easily and naturally".

Creating a brilliant future technique

This technique can be used to help clients achieve any kind of goal or outcome they want to, eg weight loss, becoming a non-smoker, overcoming phobias, giving a presentation, etc.

Here's what to do:

1 Ask your client to create an image of what their life will be like at some point in the future when they have achieved their goal/outcome.

2 Ask the client to put two pieces of paper on the floor – one labelled to represent today and one for that time in the future when they will have achieved their goal. The distance (time) between the two spots should be what feels right. They should stand on the future spot and imagine what it is like having achieved their goal. Ask them to imagine they have a remote control like the one for their television. They use it to intensify the qualities of their internal imagery and sound. Turn up the brightness, increase the colour, improve the contrast, make it bigger and bring it closer. Turn up the volume and listen to the sounds. Have the sound tuned so that there is no interference. Ask them to step into the picture and notice the feelings of satisfaction and achievement of having achieved their goal.

3 Place some sheets of paper at equal intervals between where they are standing now (future) and the sheet representing 'today'.

4 Ask your client to stand just beyond the 'future' spot and look back at the 'today' version of themselves – the one who is just starting out on their journey to achieve their goal. Ask them to notice how it feels having achieved their outcome(s)? From that position, ask them what they have to say to the today version of themself? Do they have any tips or words of advice?

5 Put down a piece of paper with the word 'Identity' on it and get the client to stand on it. Ask your client how they have changed, now they have achieved their outcomes.

What is different about them? What role are they now playing that they were not playing before?

6 Put down a piece of paper with the words 'Values & Beliefs' on it and get the client to stand on it. What values have they changed, if any, and how have their beliefs changed in order to achieve success?

7 Put down a piece of paper with the word 'Capability' on it and get the client to stand on it. Ask how their capabilities have changed? What did they learn along the way?

8 Put down a piece of paper with the word 'Behaviour' on it and get the client to stand on it. Ask what they did along the way? What are they doing differently now?

9 Put down a piece of paper with the word 'Environment' on it and get the client to stand on it. Ask what impact has achieving their outcome(s) had on those around them? Is their environment still the same or has it changed at all? If so, how?

10 Revisit the cards in any order they wish if they feel that there is still work to do. They will know when they have true alignment because they will have a burning desire to make their first move towards achieving their goal.

Seeing the sheets of paper on the floor gives your client a better concept of the time that it will take to achieve their goal.

NLP Logical Levels

See *Making Changes* in *Hypnofact 4* page 9.

In his book *Changing Belief Systems with NLP*, Robert Dilts identified six logical levels:

- Spirituality – functions to transmit God's will to us.

- Identity – defines our mission or purpose in life. Does change reflect who I am?

- Values and beliefs – give us the internal permission and motivation to change. Why make the change?

- Capabilities and skills – give us directions about how to make the change and what new resources we need to develop in order to do so. Change how?

- Behaviours – tell us which actions and reactions need to be changed. Change what?

- Environment – we need to identify the obstacles and barriers that need to be removed in order for us to make the change. Where to change?

Dilts goes on to say that the ideal is to get all levels aligned with each other.

Gregory Bateson pointed out that in the processes of learning, change, and communication there were natural hierarchies.

The function of each NLP Logical Level in the hierarchy was to organize the information below it. Changing something on a lower NLP logical level of the hierarchy could, but would not necessarily, affect the levels above it. However, making a change at an upper level would necessarily change everything below it in order to support the higher level change.

In other words: "whatever is on top runs everything underneath". So if you make a change at a lower level but the problem is at a higher NLP logical level, the change is not likely to last.

The "Disney strategy" for creativity

Robert Dilts' Disney strategy is similar to de Bono's six thinking hats technique in some ways. The technique involves using different thinking styles – realist, dreamer, critic, and outsider – looking at the issue from each perspective. You can put four cards on the floor with those words on them and ask your client to step on each one in turn and get them to tell you what they are thinking. The idea behind this technique is to help people see how they feel about a decision and explore alternatives. Firstly the client stands on 'outsider' and describes an analytical view of their choices. Then they step on 'dreamer' and brainstorm a number of alternative choices they could make. Then they step on 'realist' and evaluate each of the ideas that the dreamer came up with. They then select the best idea and come up with a plan. And, lastly, your client steps on the 'critic', and identifies strengths, weaknesses, opportunities, and threats (SWOT) with the plan in order to improve it. This gives your client a working plan of what to do next. It's important to break state between each stage. If your client hasn't come up with a plan, go back to the dreamer space, incorporating what they've learned from the other positions. They can then go through the positions again until they find a plan that's right for them. If other people are involved in the dream, discover how it is for them by stepping into their shoes. If there is anything uncomfortable for them, ask the client to go through the stages again until it works for everyone. Lastly, ask your client to step into the attractive, compelling choice they have made and experience it fully as if it is actually happening.

New Behaviour Generator (NBG)

The New Behaviour Generator is a way to build strong motivation for a new behaviour (like going to the gym).

Here's what to do:

1   Your client needs to identify a task they'd like to get done, but don't necessarily enjoy doing. It can be anything.

2   Get your client to sit back and relax, with their eyes looking up and to the right (or to the left if they're left-handed). They should visualize a person who looks just like them, close by. See that future version of themselves having just completed the task, and see how it looks with the task already done. They should notice the results of having done it, and all the benefits that arise from that (immediately and in the future).

3   Next they need to visualize their future self doing the task easily and enjoyably. They should notice that as that future them does the task, they keep looking at the image of themselves having completed the task. Every time they look at the task, they get a good feeling of anticipation, and hear positive and encouraging internal voices. They should see how good their future self feels about the progress they're making, then see them having a sense of joy and pleasure at having successfully completed the task, and enjoying the benefits.

4   Ask your client whether this version of the future looks like something they'd like to experience? If not, just allow a gently swirling whirlwind to descend over the images. The whirlwind is a symbol of transformational change, so they can allow their unconscious mind to make whatever adjustments are necessary to make the representation right for them. When their unconscious mind has made the necessary adjustments, the whirlwind will clear, and they can see the whole process for that future them.

5   Do they want to be that future version of themselves? If not, repeat steps 3 and 4, fine tuning the benefits, ensuring the various internal representations are attractive to them. Once they have a future version of themself that they're happy with, allow that future them to move into them (into their chest, their head or their whole body) bringing all that learning with them (they can use their hands to pull them in if they wish to).

6   Identify when they're next going to do the task, and see themself doing it, easily and enjoyably.

Timeline method

Here is an effective timeline and future pacing method for making all sorts of changes, eg to rid negative associations and conditioning of past experiences, a similar approach focusing solely on step 4 can be used. For a projection of achieving future goals, either step 7 can be used solely, or with the other steps.

Here's what to say to your client:

1   "Think of a few past successes. Do not get caught up in the details, just assign a name to identify each."

2   "Think of a few learning experiences (mistakes, etc). Once again just assign a name to them, and the year or age that they happened."

3   "Close your eyes. Now start visualizing going back in time to those experiences. Drift out of yourself and move towards your past, to your first past success, then on to your second, and so on. Re-live each of the experience. Feel the joy and satisfaction from each success."

4   "Now proceed to the learning experiences. Move towards each of the learning experience. As you get into it, drain all negative associations from the experience. Visualize washing away all anger, hurt, pain etc. What's left are only the learning and lessons from it."

(When doing this, position yourself above the experience so that you are looking down on it. This helps reduce the intensity, pain, and anger of the experience. It puts you in a position of control over it so that you will not get caught up in the negative feelings instead.)

5   "When all is done, drift back to the present. Look back on your past and line up all the past successes and lessons in a row just like a runway. Each success/lesson is represented by a glowing light."

6 "See yourself in the future (by looking towards the future timeline direction). Now push the runway of success and lessons into the future. See the power, beliefs, and abilities developed from the past go into the future you and become integrated."

7 "Drift towards the future timeline. See yourself in the future achieving your desired outcomes and becoming the person you desire to be. With the abilities and experience from the past integrated into the future you, see yourself being stronger than ever and achieving all goals."

8 "Drift back to the present feeling calm and refreshed. Know that your past has given you many great experiences, and that your future is looking better than ever. Open your eyes."

Timeline re-patterning

This is a technique to clear up negative emotions. (You can use a balloon for the following technique.)

Here's what to say to your client:

1 "Identify a negative attitude or a feeling you have that is holding you back. It's an attitude that's preventing you from fulfilling your potential in life, or achieving a goal, ambition, dream, living life to the best of your ability. It could possibly be an attitude that is sabotaging your confidence to have a relationship, stopping you from fulfilling your potential, preventing you from losing weight, eg you're not good enough, you're never going to succeed, you will never be happy, you're never going to succeed at that job, or a story from your past you keep recycling that is not productive."

2 a "On a piece of paper, write down some significant events that have happened in your past related to life, positive and negative (for example the first time you went to secondary school, had an argument with a family member, graduated from uni, broke up with someone you loved, bad days, and good days)."

 b "Now write done some of the things you're looking forward in the future (eg building a business, relationship, excelling at work, family, travelling to new places)."

 c "Now close your eyes relax and imagine a line that represents your positive future, placing all the exciting things you're looking forward to on a line that represents time."

 d "Now with your eyes closed imagine the present."

 e "With your eyes closed imagine those past events you wrote down, as you imagine all your past events imagine placing them on the line that represents time in your past."

 f "Now imagine joining your line of the future, present, and past events."

3 "Now remembering that negative attitude or a feeling you have, imagine going backwards in time identifying times in your past where you had that negative feeling before. At each time you notice those negative feelings, make a mental note and continue backwards to the very first experience you had of that attitude or feeling."

4   "When you reach the very first experience of that attitude or feeling, detach yourself from the feeling and look at the memory from the left hand side."

5   "See your younger self, what happened for you to feel these negative thoughts about yourself and the people around you."

6   "Next step over the timeline to the right and look at the memory from another perspective."

7   "Now imagine floating above the experience as high as you possibly can until you can barely see it."

8   "Now float back down and imagine going to a time shortly before the event that produced those feelings or attitude happened. You begin to realize those feelings were not always there. There was a time when you didn't have those negative feelings."

9   "Gather all the information about what happened to make you feel this way from different perspectives. Imagine walking or floating back alongside your timeline to the present."

10 As you arrive at the present, look back along your timeline of past events to that very first instance you experienced these negative feeling or attitudes. Determine what resources you have now at the age you are. What life experience and wisdom do you have now that would have been useful in that experience when you were younger? When you were young you did the best you could with the knowledge you had, and now you are older, you can deal with those feelings a lot better."

11 "Fully associate into resources you have now (eg confidence, intelligence, and wisdom), notice what you see, hear, and feel when you're at your best. And Imagine filling up a bag with all the positive resources you have now, and taking them back to your past."

12 "Now bring these resources back to your past. Imagine walking back alongside your timeline in a place immediately before the memory of the past. Imagine passing all these resources back to the person you were when you first had these negative feelings about yourself, passing resources such as confidence, intelligence, experience, wisdom."

13 "Having passed those resources to your younger self, how is your response different to the experience you had that made you feel inadequate or negative? How do you feel differently about yourself?"

14 "Now let those negative emotions go, let them go forever, your mind works to serve you and it may have been holding these thoughts to protect you though now it's time to let go."

15 "Now picking up your balloon, exhale all those negative feelings into a balloon, get rid of those feelings into a balloon, blow it as big as you can, and imagine getting rid of every part of those negative feelings or attitude you had. Once you are rid of the negative feelings, tie up the balloon and burst those feelings."

16 "Now make your way back to the present having released all those negative feelings, and, as you make your way back in time, each time you come across the same negative feelings, release them, and replace them with the positive

resources. As you make your way back to the present notice how you feel releasing all the negative energy."

17 "Having released all the negative energy, imagine how you will respond differently to events and interactions with people, relationships, work, and the impact it will have on your life. Having released your negative feelings, imagine two weeks into the future, then two months, four months, six months, then one year into the future laying down these new positive resources at each point."

18 "Now from the future, face the present and notice the changes you have now made with those new positive resources, letting go of the old negative ones. Give your present self whatever information you have that will assist you in making those changes."

19 "Now gradually come back into the here and now into the present, feeling positive, refreshed, recharged, focused, and determined."

Floating a goal into your timeline

Here's how to help a client achieve their goal:

1 Ensure that their representation of the goal is clear and has energy. They should step back out of it and put a frame around it.

2 They then take hold of the frame. They should be aware of their timeline and float up above it.

3 They then float forward above their timeline into the future until they are above the date when their goal will be achieved.

4 Tell them to let go of their goal and let it float down to that point on the timeline. Let it bed itself into the timeline.

5 They should notice as all the events leading back from their goal to now change and realign themselves to support their goal.

6 They should notice any changes in the future timeline out beyond their goal.

7 They should float back to above now and float back down to the present.

8 They should notice the first thing that they need to do in order to achieve their goal – and do it!

This process is good for:

1 Primarily visual people, people who need to close their eyes to visualize

2 Using with groups

3 Where there isn't room to lay the timeline out on the floor

4 Where you don't have so much time

With individuals, and where you have room, you can also use the walking a goal into your future timeline process.

Walking a goal into your timeline

Here's what to do with a client if you do have room on the floor:

1 Lay their timeline on the floor, and ask them to step onto now, facing the future.

2 Find out where in the future they want to have achieved their goal and establish where that is on the timeline.

3 Ensure that their representation of the goal is clear and has energy.

4 They should carry their goal and walk towards the future and stop just before the appropriate point in their timeline.

5 Here they can let their goal float down onto their timeline.

6 They can then step into the achieved goal and feel the experience of achieving it. This experience must use all their senses (visual, kinaesthetic, auditory, taste, and smell).

7 Ask your client to turn up the intensity.

8 Get them to step beyond their goal and look back towards now. Ask:

- What plans did you need to develop?
- What skills did you need to develop?
- What were all the things you needed to do?
- Whose help did you need to call on?
- Who did you look to model?

9 Ask them what advice they would give to themselves back at now?

10 They close their eyes and walk back to now. As they do that, they should notice the events lining up to support them in achieving their goal.

11 Get them to stand at now, take a few moments to reflect on how they feel about achieving that goal now. Ask them what is the first step they need to take right now?

Clearing anxiety

This should reduce anxiety for any future event.

Here's what to do:

1 Say to the client, "Float up above your timeline and into the future to 15 minutes after the successful completion of the event about which you thought you were anxious. Tell me when you're there."

2 "Good. Turn and look back towards now."

3 "Now, where's the anxiety?"

4 "Come back to now."

5 Test by thinking about the future event. Where is that anxiety now? Try to feel it – what happens?

Empowering questions

Write out 10 empowering questions, for example:

- How am I going to enjoy improving what I do?
- How can I become more passionate about helping others to succeed?
- What's going to be the most fun, and going to move me forward today?
- What's the most effective way I can move forward?
- Who can help me with this?
- What can I do to help others succeed?
- What am I going to stop doing today that in stopping will help me succeed?
- What's the most useful thing I can do right now?
- How can I solve this problem and have some fun doing it?
- How am I going to get a better result and enjoy the challenge?
- What is the most useful thing I can do next, and how can I become absorbed doing it?
- How have I managed to solve problems and had fun doing so?
- How have I got better results and enjoyed the challenge?
- What are the most useful things I've done and how have I become absorbed doing them?

Ask yourself these questions twice a day (perhaps first thing in the morning and last thing in the evening), pausing between each question to let the question resonate throughout your mind and body. Revise and improve your questions every two weeks or so.

Feedforward

Feedforward is an approach developed by Marshall Goldsmith in *What Got You Here Won't Get You There* to make it easier for people to continually improve what they do.

Here's what to do with your client:

1 Pick one behaviour that they would like to change that would make a significant, positive change in their life.

2 They can describe this objective in a one-to-one dialogue with anyone – it could be their wife, children, boss, best friend, or co-worker; it could even be a stranger. The person they choose is irrelevant. They don't have to be an expert on the subject or on your client.

3 They ask that person for two suggestions for the future that might help them achieve a positive change in their selective behaviour. If you're talking to someone who knows you or has worked with you in the past, the only ground rule is that there can be no mention of the past. Everything is about the future.

Their ideas represent feedforward.

4   They must listen attentively to the suggestions, taking notes if they wish. They must not judge, rate, or critique the suggestion in any way. They can't even say something positive, such as, "That's a good idea". The only response permitted is, "Thank you".

5   They can then try out the ideas.

Fool-proof planning

This is taken from Richard Bandler and Garner Thomson *TRANCE-formation*.

Here's what you need to do with a client:

1   Ask them to step into a full sensory representation of the way they will be behaving, talking, thinking, and feeling when they are completely on track with their new and preferred direction. To intensify the experience, they can imagine going through an entire 'ideal' day with their new resources already in place, building on their good feelings.

2   Ask them what needs to be done immediately before they could have their perfect day. Make a note of their answer.

3   Then ask them the same question: "what needs to be done immediately before you achieve that step?" Write down the answer.

4   Keep doing that until they have moved backward to their starting point. They should now have all the key steps needed to carry them from their present state to their desired state.

5   Carefully give each step a start and finish date, making sure that they all complete within your client's overall timescale.

Getting things done

This is also from Richard Bandler and Garner Thomson's *TRANCE-formation*.

Here's what to do with a client:

1   Get them to choose a situation where they feel out of control, not because they don't have the knowledge or the skills, but because their emotions get the better of them.

2   Help them to understand at this point that this is simply an attitude that is stopping them from doing something that they know they should do. Then they can decide what they will be doing when they are back in control. They can choose a specific example of this behaviour, preferably one that is immediately testable.

3   They need to sit comfortably, then float out, imagining themself sitting a little behind and up from their physical body. In their mind's eye, they should see the back of their head, the width of their shoulders. They should see what their clothes look like from this point of view.

4   Now, ask them to imagine that they can see themself starting to stand up, and, as that happens, your client should actually stand, so they are in precisely the same position as their imagined image.

5　Ask them to repeat this thought and action several times, making it faster each time, until they feel themself being pulled to their feet by the vividness of their image.

6　Get them to imagine they are standing a little behind and up from an image of themself about to start the activity they identified in Step 2.

7　Run the activity from start to finish several times. Do this faster and faster, stepping into the image each time, until they feel the same 'pull' as before.

8　Test by starting the activity and following it all the way through at least three times. Then they should sit down quietly for a few moments and imagine how their life will be different and better as this new skill generalizes out into other, equally useful and appropriate, areas of their life.

Gratitude inventory

This is also from Richard Bandler and Garner Thomson.

Here's what to do with your client:

1　In a notebook, ask them to draw three vertical columns. Give each column a heading: 1) Assets, 2) Attributes, and 3) Relationships.

2　Now get them to begin to fill out each in some detail.

Under assets, the might put something material (for example, "I own my house outright.") or physical (for example, "I'm the fittest I've ever been"). Assets may also refer to their experience (such as educational qualifications, work experience etc).

Under attributes, they might put practical abilities, such as being able to complete projects on time, as well as emotional capabilities such as patience, endurance, even temper etc.

Under relationships, they might list all those people (friends and family) they are confident that they can rely on. They should think: If I were in trouble in the middle of the night, could I ring any of these people and know without a moments doubt that they would turn up to help me?

3　When they have at least 20 in each category (if they don't have enough they can ask for suggestions) they should choose one from each.

4　On a separate page, one for each item, ask them to write a sentence beginning with:

"Today I'm grateful that / for...." and fill in the relevant word of phrase.

5　Each morning, get them to take a few moments to immerse themself in gratitude. This is how they do that.

Find one example of each asset, attribute, and relationship. Imagine they are actually experiencing each in turn. They should absorb themself into what they see, hear, and feel and let the physical sensation this triggers begin to flow through their body. They should notice the direction in which it is moving, and link the end point to the beginning so that it begins to spin. Now speed up the spin, allowing the feeling to move muscle to muscle, cell to cell, atom to atom, until they have a full body experience.

6　Next they should briefly imagine themself moving into the day with that feeling permeating their whole being.

Ask them, how would it change your experience? What would be different and better? Most importantly who are the people you would like to share this special kind of wealth? How will you do it?

Overcoming overwhelm

This process will assist you to identify when a client is feeling overwhelmed and lead them to a resourceful state where they can take responsibility and where positive outcomes can be set.

Here's what to do:

1　Identify a resourceful state.

2　Build rapport.

3　Identify overwhelm. Notice the client's language. This will often include generalizations such as:

- "Everything's getting on top of me."
- "My life's a shambles."
- "I feel everything's a waste of time."
- "I don't know where to start."
- "I've got so much going on in my mind I seem to keep going round in circles."
- "I feel like I'm in a deep dark hole and I don't know how to get out."
- "I can't see the wood for the trees."
- "There seems to be no light at the end of the tunnel."

These types of phrase indicate that the client is overwhelmed. The person might also just start talking, jumping from one issue to another, without stopping.

5　Pace:

Pace and stay meta by using reflective language to separate the map from the territory, eg "So, for you, at the moment, it seems like everything's getting on top of you?" Or, "It seems like there is so much going on in your life that you don't know where to start?"

6　Pre-test:

"So you'd know if the way you feel now changed, wouldn't you?"

Chunk down and get specific information.

Remember, the structure of overwhelm is to make huge chunks, which are too big to cope with. Get reasonably specific with what the issues are; for example, "£3,000 debt" is a more manageable chunk than an all-encompassing "major financial problems".

7　Pace:

An analogy is sometimes useful (eg "It feels as if a truck full of horse manure has backed up and dumped the whole lot on you!"). Humour will help to lighten the situation and lead the client to a more resourceful state that you can anchor and use later.

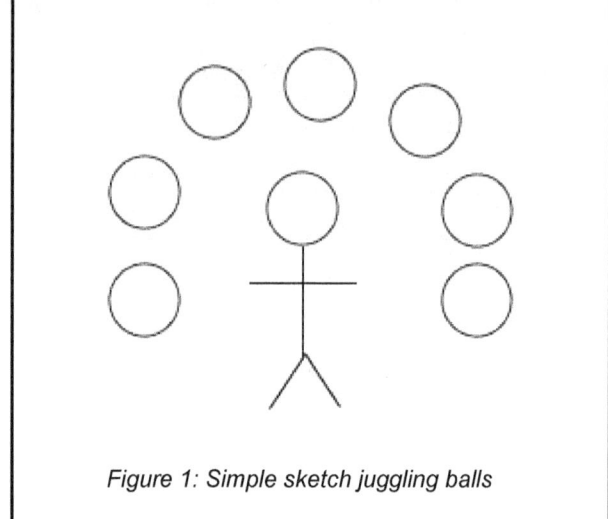

Figure 1: Simple sketch juggling balls

8 Do a simple drawing of the situation (see Figure 1):

The drawing has the effect of showing the client the big picture in smaller, manageable chunks. This effectively gets the issues outside of the client's head and allows them to be objective (dissociate – third position) about resolving them. It also helps them to see and understand why they feel the way they do. (Note: each circle will contain an 'issue'.)

9 Pace and lead:

Yes sets are useful here. For example: "So you've been feeling overwhelmed... You've got these issues here to sort out, and you haven't really known where to start... And today you've come to talk to me... and that means you've already taken the first step towards resolving them."

10 Pre-frame resolution and thinking of the larger system:

You can use other language patterns to pre-frame the situation as having various solutions. Often people experiencing overwhelm will think that the issues are so big or that there are so many of them that fixing one or two will make no difference. One way of pacing this and pre-framing a solution is: "Under normal circumstances you would be a resourceful person who would easily be able to handle any one or two of those issues. It's just that there's a whole lot happening at the same time and it's understandable that it's not so easy for you to cope. Sometimes we think that even if we fix up one issue it won't make any difference because there are so many others. However, most aspects of our lives are linked together in a system, so that when we start to fix one area, it automatically has a positive impact on some of the other areas too." You can give examples of this, relative to the issues they've described. "This means that just starting on one issue will have a positive impact on many of the other issues. And often, once one or two are resolved, the others resolve themselves, or at least they seem more manageable."

11 Ask questions:

Show them the drawing again and ask:

"Which one issue, when resolved, would have the most impact on all the others?"

"Which is the easiest one for you to resolve?"

"Which one would you like to start with?"

If dealing with the biggest issue seems overwhelming or is impractical at that point, resolving the easiest will help build confidence and reinforce the person's problem-solving abilities. Check to see which option the client prefers.

12 Tackle one:

Use the conditions of a well-formed outcome. Establish the first steps and ensure the client has the resources needed to be successful.

13 Check:

"When you think about the issues you came with, notice how you feel differently about getting them resolved now."

14 Future pace and ecology check:

"When this issue that you're going to deal with first is resolved, what will be the positive outcome on the other areas that were an issue?"

In subsequent sessions it's useful to check on the client's progress with working on the things they agreed to deal with first and help him or her establish ways of overcoming the other issues.

The relaxation response

This is taken from Richard Bandler and Garner Thomson's *Secrets of Being Happy*.

This should be used by your client for 15-20 minutes, ideally twice a day. They should:

1 Start by selecting an anchor word such as 'peace', 'calm', 'quiet', or 'relax'.

They can also choose a neutral word such as the number 'one'. Whichever word they choose, it should remain the same for all sessions from now on.

2 They need to sit quietly and comfortably for a few moments to allow their body to start to settle down.

Each time they exhale, softly and silently they repeat their word. They aren't required to visualize anything or try to relax, or do anything else. Simply think the word once each time they breathe out.

3 They must accept that thoughts and distractions will come, but they shouldn't fall into the trap of trying to resist them. Rather, each time they realize they've been thinking of something else, gently return to the word. It doesn't matter how they feel during the exercise, the benefits appear afterwards as they increase their capacity to accept whatever it is.

Success state exercises

This exercise leads your clients into great states that in turn will empower them in all aspects of their life. It's from *Conversations* by Richard Bandler and Owen Fitzpatrick.

These tasks need to be performed on an ongoing and consistent basis:

1 Awareness

Spend five minutes every day paying attention to what you see, hear, and feel in the world, noticing the differences that exist. Notice the difference between what you can see, hear, and feel on the outside and be aware of how you are thinking on the inside.

2 Openness

Spend five minutes every day taking a belief that you have about something in the world and practice looking at it from the opposite point of view. Construct as many arguments as possible thinking about it from different perspectives.

3 Curiosity

Begin to ask more questions of everything. Buy more books. Find the people who do things you really admire and ask them how they do it. Realize you can learn from everything. When you meet people from a different background to your own, find out about their culture. Ask enough questions to gain a good understanding.

4 Flexibility

Every day practice new things. Try new meals in restaurants, try different routes to and from work. Take up something you have never tried before and give it a go. At least once a week, do something that you have not done or gone to before.

5 Determination

Every time you don't get what you want, simply focus on what you want and change your approach and start again. Keep going and only focus on how good it will feel once you've done it. Avoid the possibility that it might not be achieved.

6 Pragmatism

Practise repeating the question: "What is the most useful thing to do now?" Each time you come face to face with a problem repeat this question.

7 Responsibility

Whenever a problem crops up, always ask the question: "What can I do now?". Make this a habit too.

8 Creativity

Once a week, spend 20 minutes and take three ideas that have nothing to do with each other and invent some new product or philosophy from combining them. No matter how ridiculous it seems at first, keep practising this exercise.

Spinning technique – to remove anxiety

This is a really simple and effective technique that removes any negative emotion. The technique was originally developed by Richard Bandler and further developed by Tim and Kris Hallbom.

Start by ensuring your client feels comfortable and closes their eyes. Then say:

"Now really imagine yourself in the situation that is creating that negative feeling. Just spend a few moments getting that feeling."

"Begin to notice where that feeling starts inside your body."

Take enough time for the client to locate the feeling in their bodies. If the actual sensation is only experienced in the head you can say: "That's good and where else?" It is important that vertical integration and connection between body and mind takes place during this process.

"It might start in your stomach for example and move all the way up towards your chest."

"Because it is a continuous feeling. It will be moving round and round otherwise it will be gone in an instant and it wouldn't bother you anymore."

"I want you to pay full attention to the sensation. What form does it have? What colour does it have? What direction is it spinning?"

If it is not spinning, ask: "What direction can it spin?" Wait for an answer and then say: "I want you to start spinning the sensation, faster and faster and faster in that direction".

"When it's spinning at its fastest, I want you to spin it outside of your body to a place just in front of you and below you so that you look down onto the old unwanted feeling you used to have."

"You might want you to imagine yourself pulling this feeling outside of your body. It's almost as if you have tiny little hands reaching inside you and pulling it outside."

"And I want you to imagine what colour that feeling is."

"And you can allow the spinning to slow down, slower and slower (if it hasn't done so already) until it stops and then starts spinning in the in the opposite direction faster and faster."

"And I want you to change the colour of that feeling to a different colour – perhaps white, or gold, or pink, or any other colour that you'd particularly like it to be."

"Now I want you to pull that spinning feeling back inside you and keep it spinning faster and faster in the opposite direction with a different colour. And as you try to think about the thing that was holding you back, spin it faster and faster."

"Now you can just stop and shake off that feeling. Shake your body and think of something different."

"Then think about how great it would be and how you would like to feel in that situation and imagine yourself feeling that amazing in that situation in the future."

"And try as hard as you can to get that old feeling back instead of just noticing how different you feel now. And that feels good doesn't it."

So if you find yourself having that negative feeling again just take a moment to properly notice it, spin it, and banish it.

Training, coaching, mentoring, and hypnotherapy

A look at coaching styles to see whether they can help hypnotherapists in their work.

Some of your hypnotherapy clients are going to treat you as a life trainer, a coach, or even a mentor. But what's the difference between them?

Let's take a quick look at the stages a person goes through while learning about anything. The first stage is called unconscious incompetence. This is you on your first day on the job. You don't know anything about anything, and you also don't know what you don't know! After a little while of orientation, watching and listening, and your first introduction to a work-based training course, you find that you start to achieve conscious incompetence. This is where you gradually realize just how much there is to learn. After a lot of work, you reach the stage of conscious competence. This is where someone asks you a question and you know that there is an answer. You may be able to solve their problem, but you need to concentrate to think about what to do. Or you may simply be able to look in the right part of the manual. You know what you're doing, but you do it slowly and carefully to make sure you get it right. The final stage is unconscious competence. This is where you know exactly what to do and get on and do it. You're able to do it without really needing to think about it. That's where you want to be in your career or life in general, and you probably don't need a hypnotherapist because all parts of your life are going well.

Training is usually important very early in your career in any job. It could be a video or online training, but, ideally, it would be human-led training, so that questions can be answered immediately. What it should do is explain the very basic information you need to do your job. What brings many people to see us is that there isn't any training for everyday life! After the training course, the trainee can go back to work and practice what they've learned. Then, at varying times through their career, they can attend more training courses and get more knowledge about more specific areas.

Whatever type of training a person goes on, the course contents will generally be decided beforehand with only a small amount of variable material, because it's important that each running of a training course covers the same material and anyone attending the course will have been told the same facts. And the trainer's knowledge is (hopefully) passed on to the trainee.

A mentor is usually someone with more experience of doing a particular job than the mentee. They can offer advice when they notice their mentee doing something wrong or struggling, and they can answer questions whenever the mentee has them. They're completely role-focused in that they help and answer questions about doing a particular role more successfully. Like trainers, mentors also pass on knowledge to their mentee, but in a more random and less structured way than a trainer would and only when required (meetings, typically, aren't scheduled). If they don't know an answer, mentors should know where to find the answer. Sometimes mentors are very senior people with lots of knowledge. Other times they are people only slightly further ahead in their career, who understand the kinds of question that a newer person will ask.

Many clients will treat their therapist as a mentor, expecting them to be more experienced with whatever problem has brought them to see you. The secret is that whenever they ask you a question about some aspect of their life, to throw it back to them and ask them what they think. That way you don't fall into the trap of always 'knowing' the answer to their problem. After all, they are the experts on their own lives.

A coach has a quite a different role. He or she doesn't need to know the details about how to do a job. Their role is to help the coachee make decisions about their career and/or issues at work. Coaching sessions usually occur at a fixed time, last for an agreed period of time, and have an agreed number of sessions. The sessions have a particular focus, which is usually to achieve some immediate goal. The coaching session should help the coachee to improve awareness, set and achieve goals, which, in turn, will improve their performance. Coaching aims to draw out a person's potential rather than share technical knowledge. It's reflective rather than directive, and develops a person rather than gives them information. It enables them rather than trains them.

This is clearly more like what we do. Interestingly, there are a number of successful coaching models. The most important skills that a coach needs to have are active listening, questioning, providing actionable feedback, and facilitating. Let's look at some coaching models.

The GROW model is, perhaps, the best known coaching model. GROW stands for:

- Goal setting – sets goals for the session as well as short and long-term goals.
- Reality checking or current reality – questions and explores the current situation and any conflicts.
- Options – what alternative strategies or courses of action are available and what Obstacles are in the way.
- Ws –What is to be done, When, by Whom, and establishing the Will to do it (or motivational reasoning behind the progression goal).

The GROW model was developed by a number of authors including Graham Alexander, Alan Fine, and Sir John Whitmore.

The TGROW coaching model was adapted from the GROW model by Myles Downey in his book, *Effective Coaching*. The T stands for Topic, which is what the coachee wants to address. This is bigger than the Goal.

The next acronym on our list is ACHIEVE, which was developed by Dr Sabine Dembkowski and Fiona Eldridge. ACHIEVE stands for:

- Assess the current situation – using rapport building, the use of open-ended questions, and active listening, the coach examines every aspect of the coachee's life.
- Creatively brainstorm alternatives – this gets coachees passed that feeling of being stuck.
- Hone goals – and makes sure that they are SMART. (It's an acronym standing for Specific, Measurable, Assignable, Realistic, and Time-related.)
- Initiate options – let the client make the suggestions, even if it means sitting in silence for a while.
- Evaluate options – sometimes taking a break before evaluating options can help.

- Valid action plan design – concrete steps are planned.
- Encourage momentum – provide encouragement and keep on track.

OSKAR uses solution-focused techniques and was developed by Paul Z Jackson and Mark McKergow. The process has two main thrusts – identify exceptions to bad things and doing more of the things that work. OSKAR stands for:

- Outcome – what does the coachee want from coaching/what do they want from today's session?
- Scaling – on a scale of 1 to 10, with 1 representing the worst it has ever been and 10 the preferred future, where would the coachee put the situation today? What would have to happen for them to be n+1? How would they know?
- Know how and resources – what helps the coachee to be at n? When do they already get a better outcome? What did they do to make that happen?
- Affirm and action – what's already going well? What is the next small step?
- Review – what did the coachee do that made the change happen? What effects have the changes had?

There's also the SUCCESS coaching model, where SUCCESS stands for:

- Session planning – what actions will they commit to? What challenges did they face since the previous session?
- Uplifting experiences – ensure the coachee remembers their successes.
- Charting their course – working on a goal.
- Creating opportunities – what opportunities exist or can be created to help the coachee move forward?
- Expectations and commitments – how committed is the coachee to their goal?
- Synergy – does the coachee's feelings match the goal.
- Summary – what did the client get out of the session.

There's the STEPPPA coaching model, where, it's believed that behaviour is driven by emotion, therefore, actions are motivated by how emotionally committed to a goal people feel. STEPPPA stands for:

- Subject – what does the coachee want to talk about?
- Target identification – choose a target and keep it in focus.
- Emotion – is the goal worth it, how does the coachee feel about it?
- Perception and choice – what does the goal mean to the coachee?
- Plan – what the coachee is going to do.
- Pace – measuring progress
- Adapt or act – do something to achieve a goal.

The CLEAR coaching model was devised Peter Hawkins, where CLEAR stands for:

- Contracting – establishing the outcomes the coachee wants to achieve.
- Listening – active listening to help the coachee understand their current situation and possible solutions.
- Exploring – helping the coachee to understand the impact or effect that a situation or behaviour has on their lives. And challenging the coachee to look at options for how to resolve the situation or change a behaviour.
- Action – choosing the next step.
- Review – looking at what's been achieved.

The RAPPORT model was devised by Seth M Bricklin as a way to increase emotional intelligence in executives. RAPPORT stands for:

- Relationship
- Assessment
- Provide feedback
- Plan for action
- Organize change
- Review progress
- Think ahead for growth

Dr Ron Muchnick came up with The SOLVE coaching model in his book, *Coaching: How to Solve Executive Coaching Issue*. SOLVE stand for:

- State the problem
- Observe the problem resolved
- List exceptions
- Verify the plan
- Execute.

The ARROW model is similar to GROW and was developed by Matt Somers, author of *Coaching at Work*. It adds reflection as an important part of the coaching process. ARROW stands for:

- Aims
- Reality
- Reflection
- Options
- Way Forward

And there are many other acronyms that can help in a coaching situation. It's worth looking at the coaching models to see how they help clients to achieve their goals – both short-term and long-term and see whether that can help us in the clinical situation.

Working with teams

As, more-and-more, hypnotherapists are working within organizations, they can find themselves spending whole days with teams – sub-groups within an organization comprising people doing similar work. Here's some useful information to bear in mind when working with those teams.

Non-linear thinking

If I were to ask any group of people, what their brain is for, they would probably reply that you have a brain so that you can think. It seems pretty obvious really. Without a brain, no thinking whatsoever would take place. But is that how the brain evolved? Did some creatures find one day that they had a structure that allowed them to think and it gave them an evolutionary advantage over the others? Yes and No is the answer because that structure giving them success in the survival of the fittest had to come from somewhere – and that somewhere seems to be as a way to control movement. It's like animals evolved a coxswain (using a rowing metaphor here) to ensure everyone (ie each part of the body that could effectively move) rowed in time to maximize their speed through the water and control the direction they went in. Over time, the cox evolved to become the captain of the ship. You might be interested to know that some adult sea squirts stop moving (become sessile) and digest their own brains because they don't need them anymore.

That, in part, explains why moving – a brisk walk or a trip to the gym – is good for our brains as well as our bodies. But, clearly, our brains do more than control our movements, they do allow us to think. And yet, it's very easy to not think very hard and to rely on habits. It's a bit of a trade-off in evolutionary terms. The more we think, the more food and oxygen we use up (about a fifth of all that's used in the body, apparently, gets used by the brain). So if we can get by without thinking too deeply about any decision we make – like should I eat that last doughnut or should I stick to my calorie-controlled diet – the fewer calories we burn. This energy saving was a good thing for food impoverished pre-humans, but makes no difference in the modern world, when we can simply pop to the supermarket.

But let's look at some thinking examples that don't need too much thinking about. If I can drive 30 miles on one gallon of fuel, how many gallons will I use up to drive 60 miles? The answer (and there's no prizes) is two gallons. And if I need to drive 120 miles, I need four gallons of petrol. This is an example of a straight line graph – as the distance axis increases, so does the amount of petrol used. And we tend to use this kind of straight-line thinking to solve lots of problems because it's easy and because it is so often correct.

Here's another example. If I drink one pint of beer (or glass of wine) I get 1 unit of pleasure – let's not worry too much what a unit of pleasure is, you get the idea I'm sure. So, as I'm finishing off that first beer and someone suggests having another one, using my straight-line thinking, I would assume that two beers would give me two units of pleasure. Therefore, how much more pleasure would 12 beers give me – or 24!?!? The trouble is, that once I start having a beer, my memory of those occasions when I've suffered from a terrible hangover the next day, or, worse, have spent the evening being ill into the toilet, disappear, to be replaced by memories of me dancing incredibly well, and being the life and soul of the party. Remember, I said they were my memories, not necessarily shared by anyone else who was there at the time!

But not all graphs are straight-line graphs. Let's suppose that I wanted to measure the height of all the adult males in Chippenham. And I then plotted the results on a graph with height running along the bottom of the graph (the x axis) and number of people going up the side axis (the y axis). The result would be (I expect) most people around the average height and fewer people being taller or shorter than the average. What we have is a bell-shaped curve (http://www.statisticshowto.com/bell-curve/). It doesn't have to always go up in the middle – it could start high at both ends and reach a low point in the middle. And it's not always symmetrical, it can be skewed to the right or the left.

What I'm suggesting is that many things in the world (like the best price to sell a product – too low and you don't make any profit, too high and you don't sell any of your product) follow a bell-shaped curve rather than a straight line. It's very easy to think of things in terms of straight lines, even when that's not how things work. And when you listen to members of these teams you're dealing with (or even your own clients) you'll hear them explaining things using straight-line thinking because it's easier rather than the more common bell-shaped curve method.

There's obviously an optimal point on the curve to determine the best price to sell any product in a neighbourhood. Encourage your team members (and your clients) to use their brains and get away from simplistic straight-line thinking to give themselves a more accurate view of the world around them and a better way of predicting what will happen next in any situation.

How to improve your skill level

One thing that your team members (and your clients) will be looking to do from time-to-time is improve their skills – to be able to do more things or to do things faster. So what are the best ways of improving our skills? Go to a conference? Attend a webinar? Read a book? What about if I told you the answer is to have a nap!?!?

If you like, you can do an experiment to show how you learn in your sleep. So, just before you go bed, do this experiment with a friend, who'll do some counting for you. Place the fingers of your non-dominant hand (I'm assuming it's your left hand for ease of explaining the experimental procedure) on the keys D, F, G, and H. Then, only moving your fingers and without lifting your hand up, tap out the sequence H, D, G, F, H, as many times as possible in 30 seconds. Your friend can time you and count how many accurate sequences you actually typed.

Of course, you could use your computer to time you. And you could type the sequence into Word or whatever and go back and count how many correct sequences you typed. The experiment just needs you to come up with a number and then go to bed and sleep.

In the morning, you can do the experiment again, typing out the sequence as before. What experimenters have noticed is that people improve on the task by 20 percent. For those of you who like to know a bit more about the original experiment, people were tested at 10pm and then again at 10am after they'd slept. Another group were tested first at 10am and then again at 10pm – so there was also a 12 hour gap between testing – and they didn't show any improvement.

This same process also works for learning facts. A 2013 review of student's revision techniques found that one of the most effective ways of learning was to spread the material to be learned over a few days rather than staying up all night cramming. This allows the brain to 'rehearse' the material being learned during periods of sleep and so the person learning the material learns it more thoroughly and retains the information for longer.

Inside the brain, learning anything, whether that's a tapping sequence or whatever, makes neurons fire at the same time, and this strengthens their connection. When you sleep, the brain replays the patterns of neurons firing that occurred during the day – and so it reinforces the learning that occurred. That way you get two lots of learning for the price of going to sleep. Sounds like a great idea!

It also means that you can practice a skill – like drumming, playing the guitar, or playing basketball – in your mind, and it has the same effect as physically performing the skill. It means that on snowy wintry evening, you can imagine yourself surfing, and you'll improve your skill. Or late at night, you can imagine yourself drumming – and you won't have any angry neighbours!

Modelling

How can any new member of the team you're working with get up-to-speed with what the team does compared to someone who may have spent 40 years working as part of that team? Well, one answer is modelling.

The thing to remember is that older staff have probably made lots of mistakes in their career. They have probably tried different things that they later found didn't work, and now they're able to get tasks done quickly and easily because they only ever do things that they know work. They've been there, done that, and got the T-shirt. Your new recruit could do the same as more senior staff and, over the course of a number of years, with a lot of trial and error, get to be as experienced as the senior staff. Or, they could simply just do what the senior staff do. They could model their behaviour on the senior staff members' behaviour. Obviously, not all their behaviour patterns, just those that are related to what happens at work in relation to the work the team performs. These newcomers will quickly get to be as effective as the senior staff are. And by quickly, I mean over months rather than years.

Where's the evidence, I hear you say? Firstly, there was the work of Albert Bandura in the 1970s. Children would go into a playroom and play with toys including a large friendly looking doll called Bobo. A little later, an adult would come in and start playing on their own. After a while they would play with Bobo, either being nice to the clown or hitting it fairly hard. Once the adult left the room, the behaviour of the child was observed – and they would usually imitate the adult's behaviour. So having a good role model is a great way to learn quickly how to behave (eg what to do in a crisis).

Also in the 1990s, Italian researcher Giacomo Rizzolatti and his colleagues were studying electrical activity in the brains of macaque monkeys. They wanted to observe the activity in a monkey's brain when it performed certain tasks like getting a peanut out of its pod – which is what they were doing. The researchers only expected to see this brain activity when the monkey moved – when it performed the action. What they actually found was that the brain activity also occurred when the monkeys were sitting perfectly still. Then they realized that the brain activity occurred when the stationary monkeys saw another monkey open a peanut pod and eat the kernels inside. The researchers had discovered mirror neurons, which fire when you see someone perform an action.

The most compelling evidence for modelling comes from 1984. The US army took four days to train soldiers to shoot to marksman level. The training they gave resulted in a 70 percent pass rate. The army decided that they wanted a way to train marksmen that took less time and had a higher pass rate. They called on Dr Wyatt Woodsmall and his colleagues. What he did was to use experts (who like any more experienced colleagues had learned what to do the hard way). The untrained staff modelled themselves on these marksmen and did what they did. This new training method took 12 hours (as opposed to 27 hours) and resulted in all of the trainees qualifying as marksmen. In fact 25 percent were graded as Expert.

So, modelling is a great way of getting new team members up-to-speed quickly.

Making the right choices

It may seem very strange, but making choices in life can be compared to buying burgers! It's an interesting analogy – just follow along with me and see whether it works for you in terms of how you make choices that meet your present and future needs.

Let's take a look at the first of these burger choices. It's very late at night – or, probably, it's very early in the morning – as you stagger out of the nightclub with your companions who are all a little the worse for wear. Maybe those shots early on were too much, or was it the Jägerbombs? And as you stroll across the car park, you can see a burger van, and you can smell the great smell of cooking meat. Realizing how hungry you are, you find some change in your pocket and treat yourself to this late-night culinary delight. But as you bite into your burger, you realize that not only does it taste awful, it's also very unhealthy for you. To put that in other words, the burger is bad now (in your current environment) and it's also not doing you any good in the future.

Let's take a look at another type of burger. It's the weekend and you find yourself at an organic restaurant where everything is healthy and nutritious and good for you. You decide to stay for a spot of lunch and cast your eye over the menu. And being a healthy environment, there are no chips, doughnuts, cakes, or chocolate on the menu. But you notice that there is a burger option and so you plump for that one. When it arrives, it's mashed up beans and comes with plenty of salad. While you know this is something that you could come to like in time, for now, you're very disappointed with the taste, but at least it's doing you good – so it's currently bad, but good for the future.

Or here's another scenario. You've been shopping and suddenly you're feeling a little peckish and you're standing very near your favourite (although not often visited these days) fast food restaurant. You go in and order your burger (and fries and a shake, probably). The burger tastes great, but you know it's really unhealthy for you. So, it's currently good, but bad for the future.

And in our fourth scenario, you make a burger at home. You have some friends coming round for a BBQ. You have some organic beef (no harmful chemicals and antibiotics for you) and you make your own burgers. You grill them and serve them in granary buns with a crispy organic salad. The burger tastes great now, and is healthy for the future.

So let's apply these burger analogies to making choices in life. Outside the nightclub burger people are miserable now and aren't working towards anything positive in the future. Farm shop burger people are miserable now, but are working towards a goal. Fast food burger people live for the moment and don't worry that they aren't building any foundations for the future. And home-made burger people are enjoying their lives now and building an exciting future.

So when you talk to your clients or the members of the team that you're working with, what kind of burger would they choose? It does give you a bit of an insight into their way of thinking. Ideally, when they make a choice, they would be choosing the best metaphorical burgers they can get for the present and the future.

Resilience

Sometimes, working in any organization can be a bit of a struggle. Sometimes members of the team that you're working with can feel out-of-step with the rest of the business. To deal with all that stress means people need to be resilient.

Confucius, that well-known Chinese teacher, editor, politician, and philosopher who lived over 2000 years ago (551 BCE to 479 BCE), is quoted as saying: "Our greatest glory is not in never falling, but in rising every time we fall". And that's the kind of resilience I'm talking about – the ability to sit through a meeting explaining an idea to a group of people who don't share your view and seem unwilling to examine it; and then bounce back so you're able to do the same thing again at the next meeting.

When we talk about resilience, what we're referring to is the ability to keep going in the face of difficulty, to spring back from adversity, and to manage our negative emotions more effectively, rather than letting them drag us into a downward spiral. Resilience is a skill that can be learned. In fact, US soldiers are being trained in resilience techniques. Resilient people:

- Are more likely to perceive challenges and setbacks as manageable
- Have greater emotional stability
- Are better able to cope with daily hassles and major stressors
- Have greater energy
- Are curious and open to new experiences
- Are good at helping others to feel good.

According to *Psychology Today*, emotionally resilient people:

- Know their boundaries – they understand that there is a separation between who they are at their core and the cause of their temporary suffering.
- Keep good company – they tend to seek out and surround themselves with other resilient people.

- Cultivate self-awareness – so they know what they need, what they don't need, and when it's time to reach out for some extra help.
- Practice acceptance – they understand that stress/pain is a part of living that ebbs and flows.
- Can sit in silence – in a sort of mindfulness way.
- Don't have to have all the answers – they find strength in knowing that it's OK to not have it all figured out.
- Have a menu of self-care habits – they have a mental list of good habits that support them when they need it most.
- Enlist their team – they know how to reach out for help.
- Consider the possibilities – they ask which parts of their current story are permanent and which can possibly change.
- Get out of their head – they can take their thoughts out of their head and put them onto paper.

Being more resilient means that you don't break down in tears at the first sign of difficulty. And that's something that many of our clients and the team members we get involved with can benefit from.

Teamwork – not always as good as you thought!

Everywhere you look, articles are extolling the virtues of teamwork – using the analogy of a football team that passes the ball to each other in order to score goals. And the same team can work together to stop the other team scoring. Wouldn't it be great, they say, if we could all work together to succeed with our goals? And then there are those team-building days when people spend time with the other members of their team. These do have the advantage of helping individuals meet other individuals and build some sort of relationship with them. Not all members of the same team work together in one office; and quite often these days, people find themselves being part of a virtual team. Team-building days are meant to boost morale and increase productivity in the workplace – I wonder what you think of them.

And so the meme continues that teams work better than rooms full of individuals, but do all teams work better than groups of individuals? Are they always good? You'll come across teams that underperform, where the expected synergy just doesn't happen. Let's have a look at the evidence why teams don't always work.

The first reason why teams don't work is because they don't know what their purpose is. Clearly, if the team doesn't know what its purpose is, its members can never know whether they've achieved it. So, if you want a group of people to work as a team, it's important that they know what their key objectives are, and what methods are being used to measure and assess whether those objectives are being met.

The second reason usually given for teams not performing well is a lack of leadership. This then leads to the team members lacking a clear and cohesive direction. While a good leader won't make all the decisions for all the members of the team, the team will know which goals they are working towards and will be able to make appropriate decisions themselves.

Often team sizes can get quite big, which leads to the third problem of there being too many cooks. J Richard Hackman, a Harvard professor and a leading researcher in team dynamics, recommends that team sizes should stay in single digits.

The fourth issue is inconsistent membership. It's been found that having a small group of people who've worked on a project for a period of time work best. Having a rotating membership by inviting some people for some meetings impairs performance. The members of a team need to know that they are the members of that team.

The fifth factor that impairs team performance is everyone agreeing! After the group has been through Bruce Tuckman's forming, storming, norming, and performing stages, they often reach the conforming stage, where everyone agrees with everyone else. It appears that having a 'deviant' in the team is really valuable. This deviant team role can help the team by challenging the tendency to want homogeneity. And by doing so, they can enhance creativity and learning.

So, to get the most out of the team you're working with, you need to ensure that the team has no more people in it than you have fingers on your hand. You need to make sure that people in a team know what the purpose of the team is – their objectives should be SMART targets (that way you have something to measure). This may result in members of the team changing – which isn't a bad thing at the beginning, but that team then needs to stay together for a while for positive improvements in team performance to be measurable. Those teams need to have a good leader who can set the direction for the team. And someone in the team needs to be prepared to shake things up from time-to-time, to get everyone to re-evaluate what they are doing and why – and so continue the improvement in the performance of the team.

Getting the most from your team

Nowadays, every department in an organization is called a team, and individuals are expected to be team players – even though there is little by way of definition of what makes a team outside of the sporting context. Notwithstanding, not only are the members of staff you're working with now part of a team, and possibly part of a larger team in the organization comprising other smaller teams, they are often sent on team-building activities. So, what can they expect if they go on these team building days?

Before we look at some of the activities you might engage in, let's look at five reasons Patrick Lencioni suggests a team might not work – what he calls dysfunctions:

- Inattention to results – the role of the leader is to keep the team focused on the results or outcomes of what they do.
- Avoidance of accountability – every team member has to be accountable for their actions and the team has to confront difficult issues.
- Lack of commitment – it's important that every team member shares the vision or buys into it. That way it's possible to gain clarity and work on getting closure for any issues that arise.

- Fear of conflict – teams become dysfunctional when they are unable to productively deal with conflict. Debate and discussion within a team are vitally important.
- Absence of trust – the ability to be open and vulnerable is a major factor in building an effective team.

Are you sure that the team you're working with isn't dysfunctional?

In order to overcome these issues, and to build the positive skills of listening, including the views of everyone, planning, being flexible enough to modify a plan that's not working, working cooperatively, and all the other skills needed for a team to work optimally, you will play a number of games during your team building day.

There's passing the ball. Let's suppose that you all stand in a circle and you're asked to pass the ball so that everyone touches the ball once, and you have to name the person you are going to pass the ball to. So people throw the ball round the circle – and occasionally drop the ball. At the end, the facilitator asks how it could be done faster. The answer is for people to move so that they are standing next to the person who threw the ball to them, with the person they passed it on to next to them. Wow, great team work. A little bit of thinking and movement and the process is so much better. Or, how about one person holds the ball and everyone else crowds round and touches the ball at the same time? An even faster result. This is how teams work by sharing ideas and best practice. Nowhere in the instructions did it say that the ball had to be thrown. That was a mistaken assumption. Watch out for those when planning work processes.

Another 'game' you might have is passing a rope through a string network of holes without touching the string in any way. Basically, a dozen holes are created by winding a piece of string around itself between two uprights. Effectively you get two rows of asymmetrical gaps between the string. The rope has to be fed through the holes in a random order. One solution is to have people carefully feeding the rope through the holes, and other people in the team holding the rope on either side of the string, so that gravity doesn't cause it to touch the string. Natural leaders come up with the plan and others make suggestions how to improve it. Definitely one where the plan can change based on experience.

A third 'game' that you might play is 'crossing the chasm'. Here you have two ropes or other markers a long way apart and a team of people with a few (the number depends on the number of individuals in the team) smallish squares of material (rubber works well). The team has to stay in contact with these squares at all times or else the facilitators take them away. The team has to put down a square and the first person has to step on it. They then place the next square in front of them and step on it. And repeat the process until all the squares are placed. At least one person has to be on each square. Then the people have to move forward until the last person is on the last square. They have to be able to step off the square while still being in contact with it. They then pick up that square and pass it forward to the person at the front, who then puts it down and steps on it. Everyone moves forward – without falling in the chasm. This is repeated until the whole team reaches the other side. Again, planning is important, working together and supporting each other are important, and listening to what people say definitely helps.

A fourth 'game' that people play is moving a nuclear bomb to a place of safety! OK, not an actual nuclear bomb, more a bottle of water. A metal ring has ropes tied to it. Some of these ropes go straight out from the metal ring and some go across the metal ring and out the other side. On top of this ring goes a metal rectangle, and on top of that goes the bomb – the water bottle. The team can't get closer to the bomb than a yard. As people start to pull on the strings, the bottle falls off the rectangle. Some of the ropes hold the rectangle tight, while others (the ones that go across the metal ring) tend to move the rectangle and make the bottle fall off. The puzzle is to lift the bottle (and the rectangle and ring) and carry it to a place of safety. It can be done. A team I was in spent time wrapping ropes round the bottle, meaning that it was held on the rectangle while everything was moved. This definitely required everyone to work together to complete the game successfully.

A fifth 'game' involved everyone putting their hands in rope handcuffs – a piece of rope with loops at both ends. But before that was done, the rope was put through the handcuffs of the people on either side of you – producing a circle of people tied together. No amount of stepping over ropes or twisting round made any difference. If you don't want to know the solution (and this is a very old rope trick) look away now! The first step is to move the other person's rope so that it is lying on your arm. It mustn't be wrapped around your rope, it should just be touching your arm nearer your elbow near your handcuff loop. Next, you feed a loop of their rope through your handcuff. Their loop goes up over your thumb and then over all your fingers. So that when they pull on their rope, the loop pulls down past your fingers and through your handcuff again. Their rope will no longer be attached to yours in any way. This process then has to be repeated for each member of the team. This exercise requires some cooperation between team members, but is all about sharing information and best practice to achieve a goal.

References:

Ben Ambridge. Psy-Q: A Mind-Bending Miscellany Of Everyday Psychology. Profile Books. 978-1781252116

https://www.psychologytoday.com/blog/design-your-path/201305/10-traits-emotionally-resilient-people

http://teambuildingireland.com/why-teams-dont-work/

Learning from the psychology of leadership

Let's dip our toes into the world of leadership theory and conflict resolution and see what we, as hypnotherapists, can learn from it.

Being a hypnotherapist is not the same as being a manager in a busy office, but there are times when you can see your client looking to you for advice and leadership. Most of the time this is not something you want to do – after all, the client knows their own life best, and they know what solutions to their problems, what goals, are best for them. But even so, they are coming into your consulting room – and because they know nothing about hypnotherapy and they have entered your space, they will be expecting you to take the lead. Especially on that first visit for their initial consultation, they will expect you to lead the session. So let's take a look at some of those management and leadership styles.

Probably the oldest way of categorizing leadership styles was by Kurt Lewin in the 1930s. He suggested there were three major styles of leadership:

- Autocratic (authoritarian) leaders, who make decisions without consulting their team members, even when their input would be useful. They use direct supervision as a way of maintaining the environment. This may create a climate of fear, where there is little or no room for dialogue and where complaining may be considered futile. This style can be demoralizing, and it can lead to high levels of absenteeism and staff turnover.

- Democratic (participative) leaders, who make the final decisions, but who include team members in the decision-making process. They encourage creativity, and people are often highly engaged in projects and decisions. Team members usually have high job satisfaction and high productivity. This leadership style is usually the most effective, although it isn't so good when quick decisions need to be made.

- Laissez-faire (delegative) leaders, who let team members have lots of freedom in how they do their work, and how they set their deadlines, although they do offer support with resources and advice if needed, but otherwise they don't get involved. This style works well with highly-qualified experts, but it can lead to poorly defined roles and a lack of motivation.

None of those seem to fit how a hypnotherapist acts. We may be a bit bossy encouraging people to listen to our download or CD. And we may be a little firm when encouraging people to do whatever homework we've suggested. But, most of the time we are trying to encourage our client, to draw things out from them. What else is there?

Back in 1964, Robert R Blake and Jane Mouton came up with a leadership style model that's called the managerial grid model. The grid is based on two behavioural dimensions:

- Concern for people – this is the degree to which a leader considers the needs of team members, their interests, and areas of personal development when deciding how best to accomplish a task.

- Concern for results – this is the degree to which a leader emphasizes concrete objectives, organizational efficiency, and high productivity when deciding how best to accomplish a task.

Originally, Blake and Mouton identified five different leadership styles on their grid. Since then, two other leadership styles have been added to the list. Let's have a look at those styles:

- The impoverished (now called indifferent) style scores low for both concern about people and for concern about production. This kind of manager preserves their job and job seniority, protecting themselves by avoiding getting into trouble. This kind of manager is concerned with not being held responsible for any mistakes, and so decisions are less innovative.

- The country club (now called accommodating) style has a very high concern for people and is quite low on concern for production. This type of manager pays lots of attention to the security and comfort of the employees, in hopes that this will increase performance. This leads to a friendly atmosphere, but not one that is very productive.

- The produce or perish (now called dictatorial or authority compliance) style is high on concern for production, and very low on concern for people. Managers provide employees with money and expect performance in return. Using rules and punishments, managers pressure their employees to achieve the company's goals.

- The middle-of-the-road (now called *status quo*) style is in the middle in terms of concern for people and concern for production. The idea is that by giving some concern to both people and production, staff will achieve a suitable level of performance.

- The team style (now called sound style) is high in concern for people and high in concern for production. Managers encourage teamwork and commitment among employees. For this method to work, employees need to feel themselves to be constructive parts of the company.

- The newer opportunistic style is used by managers who change how much concern they have for people and production in order to obtain the greatest personal benefit.

- Lastly, the newer paternalistic style changes from high concern for people and low concern for production to low concern for people and high concern for production. Managers using this style praise and support, but discourage challenges to their thinking.

This grid theory breaks down behaviour into seven key elements. They are:

- Initiative – taking action, driving, and supporting
- Enquiry – questioning, researching, and verifying understanding
- Advocacy – expressing convictions and championing ideas
- Decision making – evaluating resources, choices, and consequences
- Conflict resolution – confronting and resolving disagreements

- Resilience – dealing with problems, setbacks, and failures
- Critique – delivering objective and candid feedback.

This one is quite interesting because, obviously, we are thinking about clients and so we would expect to be high on the 'people' axis. But we are also there to do a job, to help the client achieve their goals. I'm not sure how goal-oriented our work should be. I don't think we would be like 'country club' leaders, but nor would we be 'team' style – we're, perhaps not that goal focused. What do you think?

The path-goal theory was first introduced by Martin Evans (1970) and then further developed by Robert House in 1971. You can think of it as a process in which leaders select specific behaviours that are best suited to the needs of their staff, the task they're performing, and the working environment, so staff can achieve their goals. A good leader will identify the best leadership approach to use, and then use it.

To maximize performance, a leader will:

- Help staff identify and achieve their goals.
- Clear away obstacles, thereby improving performance.
- Offer appropriate rewards along the way.

Clearly, we do help clients to identify and achieve their goals, and we do offer rewards in terms of praise for small successes. Also, we do modify our behaviour to bring out the best from each client. So that one is quite interesting.

House and Mitchell (1974) defined four types of leadership style:

- Directive – the leader tells staff what to do, how to perform a task, and schedules and coordinates work.
- Supportive – the leader makes work pleasant for the workers by showing concern for them and by being friendly and approachable.
- Participative – the leader consults staff before making a decision on how to proceed.
- Achievement – the leader sets challenging goals for staff, expects them to perform at their highest level, and shows confidence in their ability to meet this expectation.

You might think we're 'supportive' or you might argue that we use the 'achievement' style – after all, the agreed goals are challenging (that's why the client is seeing us), and we do expect them to be able to perform to their highest level!

In his book, *Leadership* (1978), James MacGregor Burns introduced the idea of transformational leadership, which is a style of leadership where the leader works with staff to identify the needed change, creating a vision to guide the change through inspiration, and executing the change in tandem with committed members of the group. Bernard M Bass (1985) extended the ideas. He suggested that transformational leadership encompasses several different aspects, including:

- Emphasizing intrinsic motivation and positive development of followers
- Raising awareness of moral standards

- Highlighting important priorities
- Fostering higher moral maturity in followers
- Creating an ethical climate (share values, high ethical standards)
- Encouraging followers to look beyond self-interests to the common good
- Promoting cooperation and harmony
- Using authentic, consistent means
- Using persuasive appeals based on reason
- Providing individual coaching and mentoring for followers
- Appealing to the ideals of followers
- Allowing freedom of choice for followers.

The four components of transformational leadership are:

- Idealized Influence (II) – the leader serves as an ideal role model for followers; the leader 'walks the talk', and is admired for this.
- Inspirational Motivation (IM) – leaders have the ability to inspire and motivate followers. II and IM constitute the transformational leader's charisma.
- Individualized Consideration (IC) – leaders demonstrate genuine concern for the needs and feelings of followers. This personal attention to each follower is a key element in bringing out their very best efforts.
- Intellectual Stimulation (IS) – the leader challenges followers to be innovative and creative.

We do work with clients to identify the needed change, and we do offer some mentoring and coaching. Hopefully, we do inspire and motivate clients, and we do demonstrate genuine concern for their needs and feelings. But not really a structure that fits therapy sessions.

Primal Leadership (2002), by Daniel Goleman, Richard Boyatzis, and Annie McKee, describe six emotional leadership styles that have different effects on the emotions of their staff:

- The Visionary Leader moves people towards a shared vision, telling them where to go but not how to get there. They openly share information.
- The Coaching Leader connects personal goals to organizational goals, holding long conversations that reach beyond the workplace, helping people find strengths and weaknesses and tying these to career aspirations and actions.
- The Affiliative Leader creates people connections and thus harmony within the organization. It is a very collaborative style, which focuses on emotional needs over work needs.
- The Democratic Leader acts to value inputs and commitment through participation, listening to both the bad and the good news.

- The Pace-setting Leader builds challenge and exciting goals for people, expecting excellence and often exemplifying it themselves. They identify poor performers and demand more of them.

- The Commanding Leader soothes fears and gives clear directions by their powerful stance, commanding and expecting full compliance (agreement is not needed). They need emotional self-control for success and can seem cold and distant.

Emotional Intelligence is a very useful skill for therapists. I would guess the emotional leadership style that comes closest to what we do is the 'visionary leader'.

The last leadership style on my list is from *Growing Pains* (2007) by Eric Flamholtz and Yvonne Randle, who developed the Leadership Style Matrix, which helps leaders to choose the most appropriate leadership style based on the type of task they're involved in and the people they're leading.

The Y-axis of the matrix defines the 'programmability' of the task. A programmable task has specific steps or instructions to complete. A non-programmable task needs a more creative solution. The X-axis describes the individual's capability and preference for autonomy.

So, with high programmability and low job autonomy, there are two leadership styles – autocratic and benevolent autocratic. With high programmability and high job autonomy, the leadership styles are consultative and participative. With low programmability and low job autonomy, the leadership styles are also consultative and participative. And with low programmability and high job autonomy, the leadership styles are consensus and laissez-faire.

It would seem that a therapy session has low programmability, and the work done by the client is their responsibility – they are autonomous – so the suggested leadership styles are consensus (sounds about right) or laissez-faire (which doesn't sound quite so good).

Other leadership styles are available – as they say, but these are the main ones. I just wondered whether knowing about these leadership styles might be helpful in a therapeutic situation. Maybe your natural style isn't proving to be successful with some of your clients? Then try another. Or maybe one style is working, but changing it may lead to even greater or faster changes for the better in your client.

The other area from psychology that I thought would be useful to look at is negotiation and dealing with conflict. Now, my experience of therapy sessions is that they are full of laughter and amusement, and clients benefitting from the changes they are getting from coming to see me. But there are occasions, perhaps that warrior client who is feeling a bit stuck, when knowing a bit more about how to persuade someone to change might be useful. Here's what my research turned up.

One method that some people use in industry to help resolve conflicts is the Thomas-Kilmann Conflict Mode Instrument (TKI). Kenneth W Thomas and Ralph H Kilmann introduced their Thomas-Kilmann Conflict Mode Instrument in 1974. It consists of thirty pairs of statements, and users must choose either the A or B item for each pair. According to the theory, each pair of statements is meant to be equal in social desirability. Taking the results, it's possible to plot a point on a grid. The grid has two axes. The Y axis goes from unassertive to assertive, and X axis goes from uncooperative to cooperative. Assertiveness is the extent to which a person attempts to satisfy their own concerns. Cooperativeness is the extent to which the person attempts to satisfy the other person's concerns.

According to the TKI, there are five different styles of conflict: competing (high on assertiveness, and low on cooperativeness), avoiding (low on assertiveness and cooperativeness), accommodating (low on assertiveness and high on cooperativeness), collaborating (high on assertiveness and cooperativeness), and compromising (in the middle for assertiveness and cooperativeness). Let's have a look at those in more detail:

- Competing is where an individual pursues their own concerns at the other person's expense. It's a power-oriented mode.

- Accommodating is where an individual neglects their own concerns to satisfy the concerns of the other person.

- Avoiding is where a person doesn't pursue their own concerns or those of the other individual, so the conflict isn't dealt with.

- Collaborating is where a person attempts to work with others to find a solution that fully satisfies their concerns.

- Compromising is where a person finds some expedient, mutually acceptable solution that partially satisfies both parties.

Perhaps, not a very good start. That doesn't seem to fit a therapeutic situation.

Another approach to conflict resolution is the 'Interest-Based Relational' (IBR) Approach, which was developed by Roger Fisher and William Ury in their book, *Getting to Yes* (1981). For it to be effective, everyone involved should listen actively and empathically, have a good understanding of body language, be emotionally intelligent, and understand how to employ different anger management techniques. So you and your client follow these steps:

- Make sure that good relationships are a priority. Treat the other person with respect. Do your best to be courteous, and to discuss matters constructively.

- Separate people from problems, which allows you to discuss issues without damaging relationships.

- Listen carefully to different interests. You'll get a better grasp of why people have adopted their position if you try to understand their point of view.

- Listen first, talk second. What they say might change your mind.

- Set out the 'facts' together.

- Explore options together. There may be a third position that you might reach together.

It's all very sound advice. We spend time on rapport building and maintaining good relationships. We do separate the client from their problem, so that we can discuss issues with them. We do listen. But we do decide on the metaphors we're going to use and we do set the homework.

There's also the Conflict Layer Model (Onion Model), which was described by Simon Fisher, Dekha Ibrahim Abdi, Jawed Ludin, Richard Smith, Steve Williams, and Sue Williams in their book, *Working with Conflict* (2000). The model explores human needs that are often hidden under layers of embarrassment, which leads people to conceal their deepest wants and divert their focus to something less important. What a person says he wants is the outermost layer and the starting point for negotiation and is called 'position'. The next layer is the layer of 'interest'. This is all about why we want what we say we want. And right at the centre of the onion are a person's 'needs' – what they really want. The negotiation is all about getting through the embarrassment to that layer.

Now that really does seem to be more like the sort of thing we do. How often do we find a client has come for one reason and by session 3 (or thereabouts) we find we're working on a completely different issue that has taken a while to come to light. Working with clients can often be like peeling your way through the layers of an onion.

In *The Psychology of Persuasion* (1984) Professor Robert Cialdini sets out six laws or rules governing how we influence and are influenced by others.

- The law of scarcity – items are more valuable to us when their availability is limited. Scarcity determines the value of an item.

- The law of reciprocity – if you give something to a person, they feel compelled to return the favour.

- The law of authority – we are more likely to comply with someone who is (or appears to be) an authority.

- The law of liking – we are more inclined to follow the lead of someone who is similar to us rather than someone who is dissimilar.

- The law of social proof – we view a behaviour as more likely to be correct, the more we see others performing it.

- The law of commitment and consistency – if people commit, orally or in writing, to an idea or goal, they are more likely to honour that commitment' Consistency is seen as desirable because it is associated with strength, honesty, stability, and logic.

We have a certain authority with clients because we are the experts at hypnotherapy. We are also committed and consistent. And, perhaps, we should get our clients to write down their commitment to behaving in a new way.

To be honest, when I started this exercise of looking at leadership and resolving conflicts, I thought the results would be more interesting than they are. Having said that, I still think that there are things we can learn from both when working in our therapy rooms. In many ways, the secret seems to be consistency while at the same time being flexible and able to modify your behaviour to get the best results possible for your clients.

References:

http://smallbusiness.chron.com/5-different-types-leadership-styles-17584.html

http://www.fastcompany.com/1838481/6-leadership-styles-and-when-you-should-use-them

https://en.wikipedia.org/wiki/Conflict_resolution

https://en.wikipedia.org/wiki/Leadership_style

https://en.wikipedia.org/wiki/Robert_Cialdini

https://www.cpp.com/products/tki/index.aspx

https://www.kent.ac.uk/careers/sk/persuading.htm

https://www.mindtools.com/pages/article/newLDR_73.htm

https://www.mindtools.com/pages/article/newLDR_81.htm

https://www.mindtools.com/pages/article/newLDR_84.htm

https://www.tools4management.com/article/understanding-the-conflict-layer-model/

Dealing with 'interesting times'

Here's how to help clients deal with life's problems.

There's a well-known Chinese curse that says: "May you live in interesting times". In the curse, the word 'interesting' means difficult or dangerous – certainly, the implication is that the future isn't safe or predictable. And that sounds a lot like why our clients come to see us. So how should our clients deal with tough times?

Firstly, it's worth noting that people are fairly resilient and can generally adapt to new situations. Mindfulness, which has recently become very popular, provides another strategy for dealing with distressful situations. Rather than fighting back, we ride the turbulence and keep our lives on track.

Let's unpick that a bit. When your client is experiencing difficult times, there are actually two problems. Firstly, there's the initial problem, which might be illness, not getting a pay rise, losing their job, or some other misfortune or disappointment. This is reality. The mindfulness technique is to face it and make peace with it. There's also a secondary problem, which is the mind's suffering – this is the response in your client's mind to the initial problem. It often starts with a judgement – such as 'this shouldn't be happening' – and that leads to an emotion, such as fear, anger, frustration, or hopelessness. Of course, emotions often lead to various behaviours, in this case it might be using drugs or alcohol, or some other destructive behaviour.

Life tends to have its ups and downs. In order to stay on track for their planned future, a person doesn't want to be pushed off course every time they have a 'down', which would make it harder to work their way to the next 'up'. So what can your clients do?

Mindfulness suggests that a person should attend to the emotion that they are feeling. But they should not give in to any distress that goes with the emotion. If a person doesn't accept that the emotion is real but the distress isn't, it can be very easy for that distress to become a permanent part of how a person feels. And that, in turn can reduce a person's choices that they can make in life. Very often, the thing that threatens an individual's wellbeing is what they think about an event rather than any actual sensations they feel. Too often, people will dwell on past events that haven't gone well; or, they will continually picture future scenarios where things go badly. You may recognize the thinking.

Mindfulness encourages people to stay focused on the present – and that way these negative thoughts about the past and the future are pushed into the background, where they lose their power to affect your client's thinking, emotions, and behaviour.

Interestingly, the famous Chinese curse that I mentioned at the beginning apparently only exists in English. There is no Chinese language curse quite like it.

References:

Practical Mindfulness: A step-by-step guide. DK. ISBN-13: 978-0241206546

Web sites that work

A look at how to make the most of your business Web site.

Most hypnotherapists are heavily involved with marketing, and their Web site is the biggest 'shop window' they have for potential customers into their business and their way of working. In this article I want to look at some considerations when creating a Web site.

Let me start by telling you a story and asking you a question. Let's imagine that there are two young girls playing together – Alice and Betty. Alice has a basket with a red ball in it. Betty has a box with a lid on it. Now, let's suppose that Alice goes out of the room and Betty takes the red ball from Alice's basket and puts it inside her box and shuts the lid. Now here's the question: where would Alice start to look for her ball? Most people would think that Alice would start looking where she last saw her ball (ie the basket) and then widen the search. This is an example of the Theory of Mind (ToM), which is the idea that other people have minds and can think about things separately from us.

According to Wikipedia, theory of mind is the ability to attribute mental states – beliefs, intents, desires, emotions, knowledge, etc – to oneself, and to others, and to understand that others have beliefs, desires, intentions, and perspectives that are different from one's own. Theory of mind is crucial for everyday social interactions and is used when analysing, judging, and inferring others' behaviours.

So what's that got to do with Web sites and hypnotherapists? Well, as stated above, most hypnotherapists work on their own and are responsible for the content on their Web site.

All too often, small businesses' Web sites are all about the business and how good it is and how highly trained the hypnotherapist is, and the pages are organized along the lines the business owner thinks is logical, and the language used on the pages is very technical. And those site owners wonder why ordinary people don't visit their Web sites very often – and if they do, those people don't stay very long, and don't book an initial consultation.

Why theory of mind is important is because you need to think what your potential customers/clients want from your Web site, not what you want to tell them. Their model of the world may be different from yours and so the current organization of your Web site pages may not make logical sense to them. It should. When you design your Web site, or when you review your existing Web site, you should imagine what your potential customer is thinking. When you're planning the navigation, think about it from the customer's perspective. This is called persona-based navigation. And you will have different types of customers, so you will need to plan the navigation for these different personas. The idea is to avoid friction or pain points – pages or forms where the customer isn't quite sure what to do or what information you want from them. In some cases you could possibly prepopulate forms or at least you can make it totally clear what format you want them to enter information such as dates (in fact, using drop-downs for this is so much better).

When you're building or reviewing your Web site, you need to keep in mind three things:

- What you want people to know
- What you want them to remember
- What you want them to do (calls to action).

And when you write the content for a Web page, you want to keep the pyramid model in mind. So, you have very few words at the top saying what the page is all about. That way, if the information is not relevant to the potential client, they can search elsewhere. They don't have to read lots of text before reaching that conclusion. Below that top text you have some more detailed information – which will clarify that they are on the right page and getting some information that's useful to them. And below that, you have all the details that interested customers may need. It's a bit like writing a press release. People should realize right at the start whether they are on the right page or not.

Interestingly, if you're talking to a group of people about your Web site and your company, you could use Kahoot (kahoot.com) for a bit of fun. Kahoot describes itself as a game-based learning and trivia platform used in classrooms, offices, and social settings. You can sign up and create a quiz about your hypnotherapy business. People can play along on their phones or tablets (kahoot.it). You can put the questions up on a screen and they need to be quick to press an answer. The app will tell them who is the fastest and their position in the order of speediest respondents. It's a fun and engaging way to give potential customers information about your company and what you do.

So, rather than thinking about what you need to tell customers and using the language you are familiar with, visualize your potential customers' journey through your Web site and make it as easy as possible for them to find the information they need and book an initial consultation with you.

About the author

Trevor Eddolls BA, Cert Ed, MOS MI, DHP, HPD, SFBT Sup (Hyp), CBT (Hyp), Dip NLP, Dip Mindfulness, AfSFH (Exec), CNHC Registered, UKCHO Registered is a clinical hypnotherapist and psychotherapist. He's clinical director at iTech-Ed Hypnotherapy and Head of IT and Social Media on the AfSFH (Association for Solution-Focused Hypnotherapy) Executive. Trevor is a Hypnotherapy Master Practitioner, a Solution-Focused Hypnotherapy Supervisor, and an NLP Master Practitioner. He has a diploma in CBT and a diploma in Mindfulness. He is a qualified Life Coach, has a diploma in nutrition, a diploma in paediatric hypnotherapy (Dip Hyp (paediatrics)), and diploma in play therapy.

Solution-focused hypnotherapy, as its name suggests, focuses a client's attention on the solution to their problems rather than the causes. Evidence suggests that dwelling on what led to a problem can increase the client's issues, whereas focusing on solutions can dramatically reduce those issues.

Trevor is a popular blogger and presenter. He has been seeing clients and writing about hypnotherapy, CBT (Cognitive Behavioural Therapy), NLP (Neuro-Linguistic Programming), and Mindfulness techniques for around 10 years.

Before training as a hypnotherapist, Trevor worked with mainframes. He also spent many years writing books and articles, and editing well-respected technical journals about mainframe technology.

You can contact Trevor at iTech-Ed Hypnotherapy in the Wiltshire town of Chippenham.

His Web site is at www.ihypno.biz.

Facebook: facebook.com/iHypno2004

Twitter: twitter.com/iHypno2004

Instagram: instagram.com/ihypno2004

www.ingramcontent.com/pod-product-compliance
Lightning Source LLC
Chambersburg PA
CBHW070050210526

45170CB00012B/658